A History of Women in Astronomy and Space Exploration

Exploring the Trailblazers of STEM

Dale DeBakcsy

PEN & SWORD HISTORY

AN IMPRINT OF PEN & SWORD BOOKS LTD.
YORKSHIRE - PHILADELPHIA

First published in Great Britain in 2023 by
PEN AND SWORD HISTORY
An imprint of
Pen & Sword Books Ltd
Yorkshire – Philadelphia

Copyright © Dale DeBakcsy, 2023

ISBN 978 1 39904 532 2

The right of Dale DeBakcsy to be identified as Author of this work has been asserted by him in accordance with the Copyright, Designs and Patents Act 1988.

A CIP catalogue record for this book is available from the British Library.

All rights reserved. No part of this book may be reproduced or transmitted in any form or by any means, electronic or mechanical including photocopying, recording or by any information storage and retrieval system, without permission from the Publisher in writing.

Typeset in Times New Roman 11/13.5 by
SJmagic DESIGN SERVICES, India.
Printed and bound in the UK by CPI Group (UK) Ltd.

Pen & Sword Books Limited incorporates the imprints of Atlas, Archaeology, Aviation, Discovery, Family History, Fiction, History, Maritime, Military, Military Classics, Politics, Select, Transport, True Crime, Air World, Frontline Publishing, Leo Cooper, Remember When, Seaforth Publishing, The Praetorian Press, Wharncliffe Local History, Wharncliffe Transport, Wharncliffe True Crime and White Owl.

For a complete list of Pen & Sword titles please contact
PEN & SWORD BOOKS LIMITED
George House, Units 12 & 13, Beevor Street, Off Pontefract Road,
Barnsley, South Yorkshire, S71 1HN, England
E-mail: enquiries@pen-and-sword.co.uk
Website: www.pen-and-sword.co.uk

Or
PEN AND SWORD BOOKS
1950 Lawrence Rd, Havertown, PA 19083, USA
E-mail: Uspen-and-sword@casematepublishers.com
Website: www.penandswordbooks.com

Contents

A Note on Inclusion		viii
Introduction		x
Chapter 1	Queen Seondeok and the Construction of East Asia's First Astronomical Observatory	1
Chapter 2	Brief Portraits: Antiquity and the Middle Ages	3
Chapter 3	Kepler, for the People: Maria Cunitz's Urania Propitia and the Popularisation of Heliocentrism	6
Chapter 4	Maria Winkelmann and the Guilded Age of Astronomy	10
Chapter 5	Eight Comets, 2,500 Nebulae: Caroline Herschel's Century of Astronomy	15
Chapter 6	Champion of Chinese Heliocentrism: How Wang Zhenyi Went from Horseback Martial Artist to Stellar Mathematician	21
Chapter 7	Brief Portraits: The Early Modern Era	25
Chapter 8	Computing Venus: The Trailblazing Path of Maria Mitchell	42
Chapter 9	The Secrets Stars Keep: Lady Margaret Huggins, Pioneer of Spectral Photography	46
Chapter 10	Making Spectroscopy Hip: Agnes Mary Clerke at the Nerve Centre of Nineteenth-Century Astrophysics	50
Chapter 11	'The Somewhat Nerve-Wearing Experience': How Sarah Frances Whiting Changed the Course of Women's Scientific Education	54
Chapter 12	'Bordering on the Marvellous': The Astronomical Menagerie of Williamina Paton Fleming	57

Chapter 13	Brief Portraits: The Great Nineteenth-Century Explosion	62
Chapter 14	Summing the Cosmos: Henrietta Swan Leavitt and the Saga of the Cepheid Stars	77
Chapter 15	She Filled the Sky: Annie Jump Cannon, Iron Woman of Astronomy	80
Chapter 16	Hydrogen Rules the Universe: Cecilia Payne-Gaposchkin and the Composition of Stars	83
Chapter 17	Before There Was Sagan: How Helen Sawyer Hogg Brought Astronomy to the People	88
Chapter 18	She Followed the Sun: Ruby Payne-Scott, the World's First Woman Radio Astronomer	92
Chapter 19	One Life for the Sun: Hisako Koyama's Half Century of Solar Observation	98
Chapter 20	Margaret Burbridge and the Dawn of Nucleosynthesis Theory	102
Chapter 21	Beatrice Tinsley, the Birth of Galaxies, and the Ever-Expanding Universe	108
Chapter 22	Legacy Suspended: Vera Rubin and the Ongoing Saga of Dark Matter	112
Chapter 23	Of Listening and Waiting: Jill Tarter and the First Forty Years of SETI	116
Chapter 24	Nancy Grace Roman and the Birth of the Hubble Space Telescope	120
Chapter 25	Studies in Expectation: Jocelyn Bell Burnell and the Discovery of Pulsars	126
Chapter 26	Jane Luu and the Discovery of the Kuiper Belt	133

SPACE EXPLORATION

Chapter 27	The Women's Space Programme that Wasn't: The Story of the Mercury 13	138
Chapter 28	Valentina Tereshkova, the First Woman in Space	145

Contents

Chapter 29	Scheduling for Success, Preparing for Disaster: NASA Flight Controller Marianne Dyson	151
Chapter 30	First: The Astrophysics and Astronautics of Sally Ride	157
Chapter 31	Preparing for an Unknown Tomorrow: Astronaut Kathryn Sullivan and the Saving of the Hubble Space Telescope	162
Chapter 32	Microgravity, I Say Thee Nay! Space Biomedical Engineer Mamta Patel Nagaraja, The Woman Behind Women@NASA	168
Chapter 33	The Concerns of the Earth and Above: Mae Jemison's Life in Medicine and Space Travel	171
Chapter 34	Brief Portraits: The Twentieth-Century Turn	175
Selected Reading List		220
Index		224

A Note on Inclusion, or 'Why Isn't [My Favourite Person] in this Book, DALE?'

So, you've scanned through the table of contents, maybe glanced through the Brief Portraits sections, and are flabbergasted, downright puzzled, as to why X is featured in a main article, Y is only in a Brief Portrait, and Z doesn't seem to be around at all. 'What is the deal, Dale? Clearly, Y should be a main, X should only be a Brief Portrait, and Z's absence is an affront to the entire science of astronomy. You jerk.'

That last part is definitely fair, but as to Z's absence, there are one of two reasons for it, the most likely of which is that I am saving them for another volume. So, Katherine Johnson, Hypatia, and Mary Somerville, for example, who are often in histories of astronomy and space travel, will all be making their appearances in the upcoming Mathematics volume, rather than this one, while Yvonne Brill will show up in the Engineering volume, Mary Sherman Morgan in either the Engineering or Chemistry volume (I haven't quite decided yet) and Hildegard of Bingen and Rhea Seddon were already in the Medicine volume. The other potential reason, of course, is that I just missed them somehow, in which case, drop me a note on Twitter – they are probably marvellous and it will be great to learn about them, and then they can make their appearance in the second edition, if and when I get to put that out. My favourite message to receive is a recommendation for a new scientist to cover that I had not heard about previously, so please poke me if your favourite person isn't here and doesn't have a second field of accomplishment under which category I might have stuck them.

As to which people are Xs and which are Ys, if I knew how my brain picked which figures to devote full articles to and which to summarise in brief portraits, I'd share that with you, but I don't, so I can't, so if you think I've committed a grave injustice in one of my choices, I ask your indulgence, and if you know of a source that I missed that might promote a Y to an X, again, just drop me a line!

A Note on Inclusion

Now, grab a blanket and some tea, settle back, and get ready to live for a while in a time when humans were just coming to grips with the night sky above them, and making their first strides in touching the stars with their understanding and invention.

Introduction

Astronomy shares with mathematics a strong whiff of antiquity. Attempts to accurately measure and formalise the assumptions of both have been in the air since Babylonian times, and so superficially one might think that women, whose names have been associated with mathematics from its earliest formulations, might be found in equal abundance in the roll calls of astronomy.

That is, to put it mildly, not the case. We possess a small handful of figures that speak to us from the depths of antiquity, such as the Babylonian Enheduanna, the Greek Aglaonike, and the Korean Seondeok, all of whom we shall meet presently, but otherwise this era represents a yawning chasm in terms of practising women astronomers.

Following Seondeok there are vast empty centuries within which we see a twinkle here and there, as a medieval nun puts out a revised calendar (as did Herrad de Landsberg in the twelfth century) or hypothesises a system of the cosmos (as Hildegard of Bingen did), but in terms of long-term impact on the scientific process of astronomical observation and computation, there is nothing to report until early Modern times, when women started contributing substantially to the advocacy of the Copernican system, assisting in the management and research programmes of private and state observatories, and carrying out the computations that reduced astronomical observations to reliable positions and orbits. Why is this?

Two possible reasons stand out for women's relative lack of representation in astronomy until early Modern times. The first is a matter of religion. Astronomy, as it was practised in the pre-Modern era, was often studied primarily for its astrological significance. From the Babylonians through to the Middle Ages, the position of the stars and planets was given a theological weight that, say, the proof of the exterior angle theorem was not. Correspondingly, the person whose job it was to measure precisely those positions had a social cachet and earning potential that the average theoretical geometer did not. Put crudely, there was money in the stars, and if there is one constant to the history of women in science, it is that when the potential to earn money creates a profession from what was previously an informal pastime, women will be the first to be excluded from it if at all possible.

A woman mathematician could be tolerated. A woman astronomer was competition, and would be so as long as people continued to place lucrative faith in the power of the stars. Cut off from the larger astronomical professions, women were also prevented from contributing in a more informal fashion by the material restrictions placed upon them. While to some degree, as in Helen Sawyer Hogg's famous phrase, 'The stars are for everybody,' available to any person willing to wander out of doors and stay up late, in a less leisurely sense they are available to everybody possessing the proper tools of measurement. These included the armillary spheres and celestial globes in use since the first centuries BCE, which were fine products of skilled craftsmen, and large-scale measuring devices like the mural sextants of the medieval era, which required access to the manual labour of others to construct.

The entrance of women in astronomy, then, had to wait for the reappropriation of the night sky from its astrological hawkers via the arrival of the gentleman astronomers from the sixteenth to the eighteenth centuries. These were men who, by and large, privately financed the construction of telescopes in their own homes and on their own estates, and who brought members of their household into their passion for observing the night sky, at first as mere recorders of their observations but increasingly as performers of complex computations and observers in their own right. While Caroline Herschel stands as the reigning example of that astronomical helpmate model, we could include in this tradition Marie Lalande, Louise du Pierry and Nicole-Reine Lepaute (who produced astronomical tables and calculations for Joseph Lalande); Sophie Brahe (who assisted and sometimes stood in for her famous brother Tycho Brahe) and Elisabetha Hevelius (who aided her husband Johannes Hevelius with observations and edited his scientific papers after his death).

As more women took up astronomy as a pastime in the eighteenth and nineteenth century, a new pool of talent entered the science, and one of their number, Maria Mitchell, became the United States' first professional astronomer, thereby inadvertently inviting the diminution of women's place in the larger astronomical community as men formed up ranks around the now-profitable profession and systematically discouraged women from acting in any capacity above that of brute force calculator and classifier. Ironically, it was this very relegation that paved the way for some of the greatest triumphs in the history of science, as the likes of Henrietta Leavitt-Swan and Cecilia Payne-Gaposchkin answered galactic-scale questions with the aid of the data dumped dismissively in their laps by their academic superiors.

Since the early twentieth century, women have made contributions to astronomical knowledge far out of proportion to their representation in that community. While, according to the American Institute of Physics, in 2017 only 33 per cent of astronomy bachelor's degrees were given to women, and in 2014 women only made up 19 per cent of astronomy faculty members, it is difficult to find an area of astronomy in which women have not made foundational discoveries. The composition of stars, the evolution of galaxies, the existence of dark matter, the nature of pulsars, the means to measure interstellar distances, the interpretation of stellar spectra, the theory of extraterrestrial signalling, the measurement of interstellar neutrinos, the development of cutting-edge orbital observation platforms – these represent but a few contributions from a segment of the population that has been given full access to the tools of astronomy for something just under a century, but has influenced how humanity views its place in the cosmos for all time.

Chapter 1

Queen Seondeok and the Construction of East Asia's First Astronomical Observatory

It is one of the great stories in the Korean royal tradition. A young princess named Deokman is brought a painting of peonies by her father, King Jinpyeong, along with some seeds of the flower. She regards the picture and remarks that the flowers are beautiful, but it is a shame that they have no scent. The king has the seeds planted, and when they bloom, the people of the court are amazed to find that the princess's prediction had come true, and the flowers possessed no aroma. The king, taken aback, asks her how she knew this would be the case, to which she responds that there are no insects on or around the flowers in the picture, which indicates that the peonies must not have a strong smell.

As with many tales from antiquity, and particularly those pertaining to future monarchs, we can be healthily sceptical about whether these events actually happened. However, here it is not so much the factuality of the events that is important, but what we are told about people's lingering historical perception of Princess Deokman, the future Queen Seondeok (595/610–647), by the story. The reign of Seondeok was one of cultural and intellectual renaissance for the Silla Empire (which lasted nearly a thousand years, from 57 BCE to CE 935, and which managed the unification of the three main Korean Empires shortly after Seondeok's reign, in 668), and as such its ruler needed an origin story to speak to her most important traits, as perceived by her time: her keen analytic eye, her forthrightness, and above all, her intelligence.

It was this intelligence that compelled the son-less King Jinpyeong to forgo his plans of elevating his son-in-law to the throne, and to give Deokman her chance to rule as Silla's twenty-seventh monarch, and its first woman ruler. That decision was not met with universal approval, and in 631 a rebellion was planned to stop Deokman's ascension. That plot was discovered, however, and its leaders executed, paving the way in 632 for

Princess Deokman to become Queen Seondeok, and to begin a reign that would last until her death in 647.

While much of her reign was occupied with forming alliances with China (then in its Golden Age under the Tang Dynasty) to fend off the opportunistic attacks from the other two great Korean kingdoms of the era, the Goguryeo and the Baekje, she lingers in the memory of her country less for her military experiences and more for her cultural, scientific, and governance innovations. Her achievements in promoting Buddhism in her country (particularly through the construction of the towering nine-storey Hwangnyongsa temple) and developing a government that sought to aid the poor and reduce taxation are accomplishments well worth discussing, but we are here for astronomy today, and so it is to the second year of her reign that we shall turn, when she had erected Cheomseongdae, possibly the oldest dedicated observatory in East Asia, and certainly the oldest one still standing.

Cheomseongdae is built of 365 stones, one for each day of the year, and consists of twenty-seven layers, perhaps to represent Seondeok's status as Silla's twenty-seventh ruler, with a base of twelve stones, likely referring to the twelve months of the year. The top of the observatory consisted of an area where Korean astronomers could lie down to observe the night sky through one of four domes placed at the four cardinal points of the compass. Seondeok's reason for constructing it is usually said to have been for the benefit of Korea's farmers, who needed better astronomical data to plan their harvest cycles, but part of the reason might have been more personal.

When she was young, a Chinese astronomer visited Silla in an attempt to convince King Jinpyeong to adopt the Chinese calendar system. Seondeok, then still Princess Deokman, was eager to speak with this learned man on the subject of the night sky, but he categorically refused to discuss such a topic with a young woman and when, later, she predicted the duration and progression of an eclipse with a startling degree of accuracy, her ability sent him into a rage during which he blurted out, 'Astronomy is not for women!' and then proceeded to remonstrate with Jinpyeong to prevent his daughter from learning more astronomy, which advice he apparently heeded, forbidding her to continue her studies of the subject. I would like to think that, in addition to the generally benevolent policy of building Cheomseongdae for the farmers of her kingdom, there was at least a little bit of personal thrill in it, as she saw her observatory rising from the ground and realised that astronomy could in fact be for women, and soon would be.

Chapter 2

Brief Portraits
Antiquity and the Middle Ages

Enheduanna (twenty-fourth or twenty-third century BCE Babylonian)

The Babylonians were magnificent astronomers who established a series of observatories throughout their territories that, over centuries, produced records of the motion of the stars, Moon, and planets which were rigorous enough to establish accurate models for the complicated motion of planets through the night sky, and the occurrence of eclipses. Generally, the astronomers at these observatories worked in collective quasi-religious groupings wherein no single individual was given particular acclaim for the accomplishments of the group; as such, we have few individual astronomers to refer to until relatively late in the empire. One of the great exceptions to this trend, however, is the astronomer and author Enheduanna, who was the daughter of the founder of the Akkadian Empire, Sargon I.

In order to integrate his new empire into the existing Sumerian system, Sargon appointed his daughter as High Priestess of the Moon Goddess in the City, a highly respected and even divine position in the Sumerian religious order. Her responsibilities included not only those symbolic and administrative functions one would expect of a religious figurehead but also the overseeing and expansion of the astronomical studies of the empire's observatory network. Further, the compiling and calculating of the 12-month Babylonian calendar possibly took place during Enheduanna's time as High Priestess, though our earliest existing copy is from the twenty-first century BCE, 200–300 years after Enheduanna's time.

For literature fans, Enheduanna is doubly significant as the earliest named author in world history, though the works ascribed to her, such as the *Exaltation of Inanna* and the *Sumerian Temple Hymns*, are more likely a mixture of pieces she wrote, pieces written by the successors to her office, and pieces written in the name or from the perspective of the High Priestess by outside authors.

Aglaonike or Aganice of Thessaly (second or first century BCE)

Aglaonike is probably the root source of the legendary power of the 'Witches of Thessaly' to pull the Moon from the sky. Said to be the daughter of King Hegetoris of Thessaly, Aglaonike was less likely a superbeing who could control the Moon, and more likely an individual who learned the Saros cycle discovered by the ancient Babylonians: that it takes 18.029 years for the Sun, Earth, and Moon to return to the relative geometry of a given configuration and, therefore, if you know when one eclipse happened, you need merely wait that amount of time, and can be relatively sure that a new one will occur that will look roughly like the original one did. So, if you wanted to put on a very good show, you could not only tell your audience when an eclipse was going to happen but how it would unfold, which would have amazed ancient audiences like the wielding of profound and powerful magic. Aglaonike is said to have boasted of her sorceress-like power to control the Moon which, if true, was a pretty solid grift for the time.

Al 'Ijliyyah or Maryam al-Asturlabiyya (tenth century)

The first astrolabe (one of the most important instruments in ancient astronomy, which allows the measurement of an object's altitude above the horizon) built in the Islamic Empire was attributed to be constructed by Muhammad ibn Ibrahim al-Fazari (d. 796 or 806), and once this first was made, the device caught on quickly. Not only important for astronomy and navigation, astrolabes fulfilled an additional religious purpose of advising Muslims when to pray, and in what direction, when far from a city. Al 'Ijliyyah was one of her era's most celebrated astrolabe builders, active in northern Syria during the reign of the Emir of Aleppo, Sayf al-Dawla (r. 944–967), who employed her as an instrument maker on the strength of her new designs and practical improvements to the astrolabe.

Fátima de Madrid (tenth/eleventh century)

Fátima de Madrid was either one of the most exceptional women of her era or else is a fabrication, whether malicious or accidental, dating back no more than a century. I am including her in this list in the hopes that it will further stimulate historians of astronomy with more access to records from the Al-Andalus era of Islamic rule over Spain to dig deeper and maybe come up with some information one way or the other to shine light on her story. Because it is a great story, even if our oldest source for it only dates back to 1924. In that year's *Enciclopedia Universal* there is described the remarkable daughter of Maslama al-Majriti, a multifaceted genius who, among other activities, studied astronomy during the reign of Al-Hakam II (r. 961–976). Together, father and daughter were supposed to have created an improved version of Muhammad al-Khwarizmi's (currently better known as the father of modern algebra than as an astronomer) astronomical tables which they also re-localised to Cordoba, then an international centre of learning, and to have fixed some problems in Ptolemy in the prediction of eclipses. In addition to that, Fátima was held to have produced work of her own, both treatises on the astrolabe, and a volume, *Corrections of Fátima*, on mathematics and astronomy. Of course, neither of those two latter texts are currently extant as far as we know.

Certainly, women of Fátima's reported level of scientific accomplishment existed in the Islamic Empire, and in the mathematics volume of this series we shall meet one of them in the person of Sutayta Al-Mahamali. To know for sure, however, whether this particular one existed, I suppose we shall simply have to wait and hope that, just this once, the cool version of a historical problem turns out to be truer than the likely version.

Chapter 3

Kepler, for the People
Maria Cunitz's *Urania Propitia* and the Popularisation of Heliocentrism

When Johannes Kepler (1571–1630) rewrote our conception of how heavenly bodies move, by replacing the ideal and eternal circles of classical philosophy with elliptical orbits along which planets move with variable velocities, he did so in part by harnessing the power of a hot piece of mathematical technology fresh off the presses: the logarithm. Originating in 1614 in John Napier's *Mirifici Logarithmorum Canonis Descriptio*, the logarithm was recognised by some as a powerful tool for discovering new connections between the measurable quantities of nature, and an even more useful widget for simplifying the painstaking computations that were the ruin of mathematical astronomers in the early modern era.

For others, however, logarithms were little more than new-fangled mathematical interlopers, parlour tricks employed by weak-willed scientists trying to avoid good honest computational work. Mathematical conservatives such as these distrusted logarithms and, by extension, did not entirely embrace scientific work that employed them. What was needed was a volume that showed the validity of Kepler's conception of the universe, but which employed mathematics that the entire scientific community – and the larger world of scientific enthusiasts – felt comfortable with. It took a couple of decades, but that task was eventually accomplished in a little-known but important text written in 1650 by the Silesian polymath, Maria Cunitz (1610–1664).

Cunitz was probably born in 1610, in the currently Polish town of Wolow, which had by that point been passed back and forth between Bohemia, Poland and Austria for centuries in a game of central European hot potato that would only continue in the years to come. Her father was a prosperous doctor and her mother was the daughter of Anton von Scholtz, who was a sixteenth-century mathematician. Cunitz (sometimes rendered as Cunitia or Kunic) therefore had the resources and the family background to carve out a path for herself, one denied to most women of her time. Her father ensured that she had a private education that encompassed mathematics, music, poetry, medicine, history and languages (by the end of her life she was fluent in seven tongues: Polish, German, French, Italian, Latin, Greek and Hebrew).

All that education, however, nearly came to naught at the hands of her century's predilection for early marriage. Cunitz was married in 1623, probably aged only 13, to the attorney David von Gerstmann, whose death in 1626 freed her to form a much more congenial match with the physician and amateur astronomer Elias von Löwen. Together, they observed Venus and Jupiter in the late 1620s and married in 1630, at the height of the continent-wide catastrophe known as the Thirty Years War (1618–1648). While hiding out from the devastation of that conflict at the Cistercian convent of Olobok, Cunitz carried out a correspondence with some of Europe's most eminent astronomers, and gathered material for her masterpiece, *Urania Propitia*.

That work's full title, *Urania Propitia Sive Tabulae Astronomicae Mire Faciles, Vim Hypothesium Physicarum A Kepplero Proditarum Complexae; Facillimo Calculandi Compendio, Sine Ulla Logarithmorum Mentione Phenomenis Satisfacietes; Quarum usum pro tempore praesente, exacto et futuro communicat Maria Cunitia. Das ist: Neue und Langgewunschete/ leichte Astronomische Tabelln durch derer Vermittelung auss eine sonders behene Arth aller Planeten Bewegung nach der länge [indec] under andern Zufallen auss alle vergangene, gegenwertige, und kunssstige Zeiespuncten furgestellet wird*, tells us a few things about the full importance of this work. Firstly, the fact that the title is half in Latin and half in German points to an important aspect of the book, namely that it is written in both what was the academic language of the time, Latin, and in a language ordinary people could actually understand, German. It was, in fact, one of the first scientific texts written in German, and so we could place Maria Cunitz as a pioneer of the SciComm movement to make specialised technical information accessible and available to the general public.

Secondly, the phrase '*Sine Ulla Logarithmorum*' which translates as 'without any logarithms', testifies to just how tendentious logarithmic calculations still were in 1650. For some, as mentioned above, they were considered too easy, almost like cheating (the seventeenth-century equivalent of a person today using a graphing calculator to factor a quadratic equation or calculate 14 x 5), whereas to others they just seemed odd and forbidding new mathematical objects that were difficult to conceptualise. (I personally see this centuries old terror re-enacted every year in students coming into my calculus course from Algebra 2 or PreCalculus; their eyes gloss over and mouths gape open in a mute scream every time the word Log is so much as mentioned.) Cunitz's book, then, promises that there will be None Of That in *her* calculations, a labour which, had her volume been more widely published, could have gone far in making Kepler's thought more accessible, sooner.

Thirdly, we see in the phrase '*Langgewunschete/leichte Astronomische Tabelln*', or 'Long Desired and Simple Astronomical Tables', another Selling Point of *Urania Propitia*. Kepler had published his *Rudolphine Tables* in 1627, a monumental effort that occupied much of Kepler's time after the publication of *Harmonices Mundi* in 1619. The creation of the tables, grounded in the data he had inherited from Tycho Brahe's legendary observatory in 1601, was a neck-breaking process, though one made simpler by the use of logarithms, which allowed multiplication and division problems (such as occur regularly with the trigonometric quantities involved in spherical geometry) to be reduced to much easier and more accurate addition and subtraction problems. The tables predicted star and planet positions based on a heliocentric model that were accurate to within one arc minute for most objects, and even those who were sceptical about his Three Laws of Planetary Motion saw the great value in the Tables.

They were, however, complicated, not only in their use of logarithms but in the solution of Kepler's Equation, $M = E - e \sin E$. In this equation, M is equal to $2 \pi t/P$, where t is the time elapsed since the planet was closest to the Sun (its periapsis), and P is the period (how long the planet takes to complete one revolution around the Sun). E, meanwhile, is an indicator of the position of the planet, measured as an angle from the segment connecting the centre of the planet's elliptical orbit to the Sun. In other words, it is an equation relating time to position for an ellipse of eccentricity e (higher e values mean that you have a less circular shape to your orbit). This is a tricky equation to solve, if you are looking for E and know M. Enter Cunitz. She vastly simplified the process of knowing a planet's position at a given time by cutting straight to the chase, and looking at the angle between the planet, the Sun, and the periapsis, instead of Kepler's E, the angle between the planet, the centre of the ellipse, and the periapsis. That angle, called the True Anomaly, could by Cunitz's method be accurately calculated directly from M, without needing to go through the difficult process to find E, making calculating future planet positions in the Keplerian heliocentric system much simpler in theory.

Cunitz's system was not perfect, as some of the coefficients she bypassed turned out to be sufficiently significant as to produce observable errors in her tables, but by and large the increase in ease of use and prediction, combined with the accessibility to a more general audience, represented an important advance in bringing heliocentrism to a wider appreciation. Unfortunately, the small original print run of *Urania Propitia* prevented it from attaining the significance that was its due, and soon a disaster would befall Cunitz that prevented her from building on the impressive foundations she had established.

In 1656, a fire destroyed most of the city of Pitisch, where Maria and Elias had settled after the Thirty Years War, and their home was among those destroyed. All of her astronomical observations, instruments and notes were consumed in the blaze, and as far as we know she did not attempt to rebuild her work from the ashes. Her husband died in 1661, and she followed him in 1664, though her name will continue for as long as our solar system is remembered in the form of the Cunitz crater, a 49km-wide impact crater on Venus that was named after Maria by the International Astronomical Union in 1991.

Further Reading

Though it now exists in only nine remaining physical copies, *Urania Propitia* has been entirely scanned into an online digital version that is free to leaf through. For her life, Gabriella Bernardi's *The Unforgotten Sisters* (2016) is a good starting point, while for the larger context of Kepler's Rudolphine Tables I do not have a book I am wildly in love with. After the big hermetic turn in interpreting the early astronomers, books about Kepler really focused on his mystical side as sort of the central case of the Hermetic Hypothesis, to the detriment of the more technical aspects of how his calculations worked and why his methods did or did not catch on from a mathematical point of view. If I had to choose one though, I would say the Oxford *Portraits in Science* volume is an accessible work that keeps the sensationalism to a minimum.

Chapter 4

Maria Winkelmann and the Guilded Age of Astronomy

Back in the age when historians favoured hard and fast lines between different eras of world history, 1543 stood as the gold standard boundary between the Old world and the Modern one. That was the year Nicolaus Copernicus's *De Revolutionibus orbium coelestium* was published, unveiling the heliocentric model of the universe from which an entirely new, increasingly secular, notion of the cosmos would grow. As such, 1543 became the shorthand boundary between the old astronomy, which managed impressive feats of accuracy but was hampered by the dead weight of astrological and theological concepts, and the new astronomy, which followed the data wherever it led and increasingly harnessed the power of mathematical analysis to form models about *how* astronomical objects moved, leaving aside the metaphysically muddled question of *why* they did. It is an interesting story, but over the course of the twentieth century, historians came to realise that the transition into modern astronomy was less digital and more analogue than the 1543 Hypothesis implied, that many of the luminaries of the Scientific Revolution (such as Johannes Kepler and Isaac Newton) held organising beliefs that harkened back to ancient hermetic traditions while the official structures that underpinned astronomical efforts for centuries bore a closer resemblance to traditional guild structures than modern academic departments. In short, during its first centuries of development, modern astronomy was managing large-scale changes in what measurements were taken and how they were analysed, while experiencing much more gradual change in the organisations and motivations pushing those new measurements.

Few figures in the history of astronomy represent the fullness of those conflicting tensions, the pull of tradition counterbalanced by the exhilaration of revolution, like Maria Winkelmann (1670–1720). Her life coincided completely with the heady days of Prussian science's first great patchwork lunge towards modernisation, directed by a few visionary souls and carried out in the face of overbearing cultural inertia. Winkelmann was born in 1670 in Panitzsch, a Saxon town of a few dozen souls near Leipzig, one of the Holy Roman Empire's intellectual capitals at the time. Her father was a Lutheran minister who privately educated her, and passed that role on to her uncle upon his death when she was but 13 years old. The young Winkelmann was such an adept student that she was soon given the opportunity of studying astronomy under Christoph Arnold (1650–1695), an amateur astronomer who had himself studied under Johannes Hevelius's most famous student, Gottfried Kirch (1639–1710), and had gained a fair level of continental fame with his 1682 sighting of Halley's Comet, and his 1686 discovery of a new 'great comet', romantically named C/1686 R1.

Like most astronomers of his era, Arnold observed not through instruments collected at a centralised institution but rather at a home observatory. Winkelmann studied as an astronomical apprentice under him, just as Arnold's master, Gottfried Kirch, had studied in the privately run observatory of his master, Johannes Hevelius, in a tradition more representative of a medieval craft system than modern academic institutionalisation. Through Arnold, Winkelmann met Kirch, who, after the death of Hevelius in 1687, ranked as the greatest astronomer of the German tradition. He was a widower some three decades Winkelmann's senior, who stood in need of a competent assistant and home organiser, and who must have represented for Winkelmann a stable opportunity to carry on first-rank work in astronomy in spite of the limitations placed on her societally by her gender. They were married in 1692, and after some time in Leipzig and Guben moved to Berlin in 1700, where Electress Sophia Charlotte of Brandenburg was employing her influence and position to bring some spark of proto-Enlightenment ideals and institutions to the often gruff and intellectually dismal town.

In the seventeenth century, Berlin was a margraviate capital of little cultural importance, which had been decimated over the course of the Thirty Years War (1618–1648) and where the most popular form of refined gentlemanly entertainment consisted of shutting one's self in a room with a group of friends and drinking and smoking until everybody passed out from the fumes and alcohol. Following the destruction and population loss caused by the war, the Great Elector had introduced a policy of religious toleration that welcomed in talented French Huguenots, who soon made up a sizable portion of the scarred city, and brought with them a greatly needed degree of sophistication. Real change, however, had to wait for the arrival of the iron-willed Sophia Charlotte, who brought Italian opera, Baroque architecture and a spirit of scientific curiosity to the capital. In 1696, she commissioned the construction of a new observatory, which was completed by either 1706 or 1711, and throughout the 1690s she was a guiding force in the founding of Gottfried Wilhelm von Leibniz's dream project, the Berlin Academy of Sciences, which opened its doors at last in 1700.

These two institutions, the observatory and the Academy, grew into existence before Winkelmann's and Kirch's eyes, and to them fell much of the task of organising, equipping and running their astronomical efforts. The Academy received no government funding, but instead financed itself through a monopoly on the creation and sale of calendars, which work fell primarily on Winkelmann and Kirch to complete. These calendars combined astronomy, astrology and meteorology to not only provide information about astronomical phenomena like eclipses and Moon phases with predictions about seasonal temperature and weather variations, as one would expect of a general farmers' almanac, but also astrological advice as to the most cosmically favourable times to undergo major (and not so major) life events.

The calendars were a major cash generator, and as Kirch's health declined, the responsibility for assembling them fell increasingly to Winkelmann, whose fame grew throughout the first decade of the eighteenth century. In 1702, she discovered a comet of her own, though in her husband's initial report on that discovery he cut her out of the credit for it entirely, only restoring her true place as the comet's discoverer in a 1710 report to the Academy. She corresponded regularly with Leibniz, then one of the continent's most esteemed polymaths, was introduced and favourably received at court, and from 1707 to 1712 wrote three separate tracts under her own name, the first on the aurora borealis phenomenon, the second on the conjunction of Saturn and Venus, and the third on the conjunction of Jupiter and Saturn, including the potential astrological significance of those events.

Everything was going just smashingly.

Then, in July of 1710, Gottfried Kirch died. It would have been natural for Winkelmann to take over his vacated position, as she had been largely performing all of its duties in the previous years anyway, and had she been working with Kirch in his private observatory, that is probably what would have happened in the best craft practice of the assistant/wife becoming the new master. Having grown up in that tradition, she confidently proposed her own name as Kirch's successor. Prussia (which had become a kingdom in 1701) had few qualified candidates to assume the role of royal astronomer, and time was of the essence, as regular publication of the Academy calendar was such an important financial pillar of the institution. Winkelmann would have been the smartest choice – somebody who knew the job and had a reputation for timeliness, rigour and accuracy. Because of her gender, however, her name was not even brought up until she put herself forward. Leibniz, for his part, took her side as an individual of talent and genius who should be supported in her continued work, but the Academy feared that it would become the laughing stock of the civilised intellectual world were it to allow a woman to take up a leadership position.

Instead, they hired Johann Heinrich Hoffmann, who quickly fell behind on his official work, leaving both his observational and calendrical duties in a constant state of unprofessional incompletion. Winkelmann, meanwhile, picked herself up and returned to the private astronomy tradition from whence she came, working as the master astronomer at the observatory of Baron Bernhard von Krosigk, with two assistants who served under her. Here, she carried on her observations, created calendars for Breslau and Nuremberg, and supported the education of her son and daughters. With the death of Baron von Krosigk in 1714, she went on to an assistant professorship in mathematics in Breslau and spent some time in Danzig rearranging the former observatory of Johannes Hevelius. In 1716, to the relief of nearly everyone involved, Hoffmann died, and the Berlin Academy decided to award his position to Winkelmann's son, Christfried.

It was a clever decision – by hiring Christfried, the Academy knew it would also be getting Winkelmann as a virtual freebie, thereby allowing them to continue the timely publication of calendars while still having a proper male to front the operation. Maria, however, was not the same person she had been six years earlier. She had experienced life as a master astronomer and mathematics professor, and was used to her intellect being heeded and her opinions respected. When visitors came to the observatory to witness rare phenomena, she made it a habit of greeting them, explaining the instruments and astronomical theories, and answering their questions. For this, she was repeatedly reprimanded by the Academy, who warned her that she was to make her presence known as little as possible, and particularly when important guests arrived at the observatory. She was to stay quietly in the background, doing the work that the Academy needed to fund itself, and which formed a considerable chunk of its intellectual output. This she apparently refused to do, and in 1717 she was drummed out of the Academy and quit the observatory. She attempted to continue observations from her own home, but with inferior instruments and inadequate access to the skyline, she experienced little more than frustration as a result.

Winkelmann died of fever in 1720. Her son, Christfried, continued in his role as director of the observatory until his death in 1740, and her daughter, Christine, as we shall come to see, continued to be employed as a calendar preparer by the observatory, where she was given the particular responsibility of preparing a Catholic calendar to serve the needs of the newly acquired citizens of Silesia after Frederick the Great's Silesian Wars. She seemingly possessed more of a gift for self-effacement than her mother, and continued to be employed by the Academy until her death in 1782.

Maria Winkelmann lived in a scientific world where the rules were in a state of constant flux, where guild and academy, astrology and astronomy, all existed simultaneously, requiring scientists of their age to possess the gift of navigating each realm's often contradictory expectations to stay afloat. It was a taxing enough job for a man, but for a woman, who in some contexts could be intellectual master of all she surveyed, while in others she was valued as little more than a silent drudge, it was a now-exhilarating, now-humiliating and generally all but unnavigable morass that yet allowed for some glory, for those deft enough to seize it. Winkelmann succeeded along multiple intellectual fronts while remaining true to her conception of the esteem due her and her accomplishments in the face of newly arisen organisations intent on devaluing them, and will hopefully stand as much an example of perseverance and integrity to the would-be scientists of the next three centuries as she has to the last.

Further Reading

Maria Winkelmann and Gottfried Kirch show up here and there in books about early astronomy and early modern women's science, which you can stitch together to form one through narrative. Two of the best sources for Winkelmann are Londa Schiebinger's *The Mind Has No Sex? Women in the Origins of Modern Science* (1989) and Gabriella Bernardi's *The Unforgotten Sisters: Female Astronomers and Scientists Before Caroline Herschel* (2016). Meanwhile, for more insight into the era generally, Johann Westphal's classic *Leben, Studien und Schriften des Astronomen Johannes Hevelius* is readily available in reprint editions, while *Hevelius, Flamsteed and Halley: Three Contemporary Astronomers and their Mutual Relations* by Eugene MacPike is a bit harder to come by.

Chapter 5

Eight Comets, 2,500 Nebulae
Caroline Herschel's Century of Astronomy

In 2092, if there are still humans on our planet to look and to see, a comet will appear in the night sky that has not been viewed since 1939, and will not be seen again until the twenty-third century. It was first noted against the stellar background by one of the unlikeliest of people in the history of Western astronomy, Caroline Herschel (1750–1848), a small, disfigured ex-maid who spent a quarter of a century languishing in domestic misery, half a century as England's most celebrated comet hunter and astronomical data compiler, then another quarter of a century as a bitter and desperately lonely former celebrity, gloomily ruminating on the scars of the past and the short memory of the present.

The first phase of Herschel's life was spent in Hanover where, if her life was a constant thrum of Cindrellic hue, she could at least take comfort in the fact that everybody else's life was equally horrid. At the time, the king of Hanover was also the king of England, which meant any time England and France were at war, Hanover was on the menu for French invasion. If it was not currently being occupied by French soldiers, it was girding itself for a coming capture.

Herschel's father was a violinist who, to keep his family in food, took up a job as a military musician, meaning that he was hardly ever home and was dismally paid. When he was gone, it was the job of his wife, Anna, to run the house, manage the half a dozen or so children present at any given time (three of the children died before their tenth birthday, and the older sons were drafted into the army band to earn extra money with their father when still in their mid-teens), and navigate lifetime during military occupation without any help.

Or nearly without, because after Caroline survived smallpox in 1754, the resulting disfigurement meant that she could not realistically be married off, which her mother saw as a grand opportunity of turning her into the household scullery maid. Anna deliberately prevented her father from teaching her music or arithmetic, lest the girl develop skills that would allow her to leave the house and take up a career of her own. For twenty-two years, Caroline Herschel was the household servant, whose job was to do whatever anybody asked her to do. Her mother battered her with words and her older brother, the thoroughly worthless and arrogant Jacob Herschel, battered her with his fists when she did not attend him quickly enough.

She was told that she was too ugly to ever be loved, and prevented from gaining knowledge of anything that might mean a way out of a future of perpetual menial service. And then, suddenly, as if written for film, her brother William made an unexpected proposal that was to change her life. William had been in England, where his career as a musician in the holiday city of Bath had proven so successful that he thought, why not for Caroline as well? He bargained with Anna for Caroline's release, giving her money with which to buy a replacement servant in exchange for his sister's freedom.

His plan was to train Caroline as a vocal soloist, and she did in fact take lead vocals in several of his productions as music director of one of Bath's swankier houses of commercial worship, but not long into her stay he caught the astronomy bug. Instead of giving her lessons in his few spare hours, he bent over mirrors, polishing and repolishing them to produce some of the best telescopes available in Europe. He started stargazing and dragged his sister with him into his new passion. Her career as a musician would simply have to wait.

Having spent two decades in a state of constant obedience, Herschel accepted this new change of life direction with resignation. She had enjoyed her time as a promising musician, the cheer of the audience that loved her voice and did not seem to care about the marks left by the pox, but her brother wanted her to be his astronomical assistant now, so those dreams had to be put aside in exchange for days and nights of polishing, measuring and learning.

On arriving in England, William had taught her enough arithmetic to carry out the keeping of the household books and to navigate the labyrinth of British currency, all brand-new ground to the woman who had been denied even the most rudimentary of mathematical education; now she had to learn algebra and geometry as well if she was to help him in his observations, and start to make some of her own.

Their task was aided immensely by William's discovery, in 1781, of a new planet. His friends suggested that he name it after the current monarch, George III, and that in exchange he would probably receive an official position that would allow him to devote all his time to astronomy and leave music for good. So, he named the planet after Saturn 'the Georgian Sidus' and received the title of official astronomer at Windsor. (That planet was later renamed Uranus by Johann Bode, which seems like something that you ought not to be able to do, but there it is.)

Caroline and William moved to their new facilities and there, while William obsessed over the gravitational interpretation of double stars and ran back and forth to keep the king's guests entertained with telescope parties, Caroline carried out observations of her own with William's equipment, the best in Europe at the time, and found herself discovering nebulae almost anywhere she turned. Her brother took note of her success and, in 1783, the two of them embarked on a mapping of the nebulae of the Northern sky that would consume half a century and net 2,500 of the phenomena, of which previously only a hundred or so were known.

Night after night, William sat perched high in the seat of his 20-foot reflector while Caroline sat at her desk beneath the second-storey window of their home. As William swept the sky and called out verbal descriptions of what he saw, Caroline's job was to take down the data, and then determine, based on her instruments, which of the standard stars in the *British Catalogue* William was referencing for inclusion in the nebular description. That was a task made more difficult by the fact that the *British Catalogue* was based on John Flamsteed's work in the late seventeenth and early eighteenth century, and that since then the wobble of the Earth's axis had caused all of the positions to precess. So, Caroline effectively had to, on the fly, while taking down William's verbal account of what the nebula 'looked like', determine what patch of the sky he was looking at from her telescope position instruments, and then work out what that patch of sky corresponded to, eighty years earlier, in order to list the correct reference star.

Her work then carried on into the next morning, when she finalised and recorded all of the previous night's observations, producing the running list of nebulae and their positions that she curated and updated all of William's life, and beyond. When she was not playing the role of amanuensis, however, she had a chance to hop up on the flat roof of her cottage and sweep the stars herself with her smaller and more convenient telescopes. She had the map of the night sky's nebulae in the back of her brain, which meant that, when something a bit nebula-ish suddenly appeared in the night sky, she recognised it as an interloper while other astronomers assumed it was just another ordinary nebula. As a result, she became her generation's most successful comet hunter, bagging eight in the short decade or so fate allowed her to chase them, including the ball of ice that we will not be seeing again until 2092.

Those comets made her a British celebrity. A public that could not quite wrap its head around the dedication and precision required to catalogue thousands of nebulae was in a position to take notice when a woman was able to find comets with much greater success than her male colleagues. But it was her earnest work as nebular cartographer that paid the bills. In 1787, she received a yearly stipend of £50 from King George to continue assisting William, making her history's second professional female astronomer (the first, Christine Kirch, we have already met as the daughter of Maria Winkelmann, and will have more to say about soon). George III was an astronomy enthusiast, and recognised the fundamental importance of the siblings' steady work, and the investment certainly paid off.

Not only did Caroline and William chart the northern sky's nebulae and, in the process, resolve the astronomical controversy about nebular composition, but Caroline, on her own, set about the task of putting Britain's star records in solid order. For a century, astronomers had relied on Flamsteed's *British Catalogue* as their standard source for star positions and descriptions, but it was by no means a perfect reference and Caroline, who had cause to reference it all the time in her work with nebulae, knew that well. So, she decided to produce an edition of it that finally presented stellar coordinates with Flamsteed's original observational data, allowing astronomers to see fluctuations in the original observations that were omitted in the *Catalogue*, thereby providing crucial data about potential star variability, while also setting right errors that had been allowed to stand for 100 years. Her work, the *Catalogue of Stars* of 1798, was so precise that, from then to now, no subsequent corrections have been needed.

The work came at a good time, because her career as an active observer had effectively come to a close. William had married a rich widow, leaving him less time to spend all night hunched over a telescope, and while Caroline got along with his new wife, some disagreement occurred in 1797 which caused her to move away from her lodgings in the cottage that had seen the discovery of eight comets. After that, her life was a rotating door of temporary lodgings that limited her astronomical output to the deskwork variety. At the height of this disorder, William even had her moving back and forth between Bath and Slough to live in whichever house he and his family happened not to be in at the time, without apparently troubling himself too much over the disruption in her life this back and forth might cause.

But move she did. Her devotion to the brother who pulled her out of Hanoverian scullery was complete, and lasted well beyond his death. She continued collating and correcting their nebular work for publication, with two editions of 1,000 nebula followed by a third of 500 that represented their last observations together. After that third volume, William's health did not permit further observation, and Caroline's unstable lodgings kept her persistently from the telescope. But she had one more massive effort in her.

After William's death in 1822, it was up to his son, John, to finish his father's work, but unfortunately the nebular record of William and Caroline was written up entirely with reference to nearby stars, a format that was very nearly useless to other astronomers. If John was to continue with their work, Caroline would have to go through all the data, one more time, and convert it all into coordinate form. The resulting catalogue, produced in her seventh decade of life, is a masterpiece of accuracy and a towering testament to her half century as an astronomer. With it, John was able to complete the picture of the Northern sky and then, once his mother passed, to take an expedition down to Africa to do the same for the Southern.

The completion of William's task was a source of great pride and happiness to Caroline, and she needed it. With William's death she moved back to Hanover to be with what remained of her family, pitifully sending them cheque after cheque in the hopes that money would buy the familial affection she had never received from anybody apart from William. It did not, and except for two or three confidants and the odd visit from royalty to commemorate her birthday, the last twenty-six years of Caroline Herschel's life were lived in solitary frustration, reliving the many psychological wounds of her youth and growling at any advances in astronomy that moved beyond her brother's work. She had honours and financial security, but as she moved into her eighties and then nineties, she outlived everyone who cared about her, and could not believe that the young people who took an interest in being with her were not just there for the money.

She died in 1848 after a quarter of a century of regretting her return to her native country. She had arranged to be buried in a vault on top of the graves of her father and mother with a new headstone that mentioned her life and the scientific societies that she belonged to, her father's career as a musician, and not one line about her mother even being buried at that location; a final expression of disappointment capping a life that appeared heroic to everybody in Europe except the one person who actually lived it.

Further Reading

Michael Hoskin is the Herschel biographer non plus ultra. He has dedicated half a century of his own to documenting the lives of Caroline, William and John. He has edited her memoirs, which make for truly gloomy reading, and written about the family as a whole, but if you want a one-stop book for all of Caroline Herschel's triumphs and losses, you definitely want his *Caroline Herschel: Priestess of the New Heavens* (2013). The research is as impeccable as you would expect, and the tone is charming and human.

Chapter 6

Champion of Chinese Heliocentrism
How Wang Zhenyi Went from Horseback Martial Artist to Stellar Mathematician

An arrow hits a target as a 15-year-old girl on a horse goes galloping victoriously by. It is not an entirely unusual sight in late eighteenth-century China – learning equestrian and martial skills was considered as normal an accomplishment to certain upper-class women of that era as pianoforte playing and embroidery were to their European contemporaries. The girl riding that horse, however, was decidedly an unusual character, one who set herself to grapple with all of the fields of endeavour offered by her time, from mathematics to poetry to martial arts to astronomy, and made a name for herself in all of them.

That name was Wang Zhenyi (1768–1797) and had she lived longer, there is no telling what her place in the world's intellectual pantheon might have been. As it was, she managed to carve for herself a national reputation in three separate disciplines within the space of three short decades, and was one of China's leading voices advocating a heliocentric conception of the solar system. Her grandfather, Wang Zhefu, was a former governor with a library boasting seventy-five bookcases of texts, and her father was a doctor and the author of a four-volume treatise on medical prescriptions.

Brought up in modern-day Nanjing, at the age of 11 she and her father travelled to Jiling to attend her grandfather's funeral. They remained at Jiling for five years after the funeral, where Wang Zhenyi occupied herself studying from her grandfather's extensive library, taking breaks now and then to learn martial and equestrian skills under the instruction of the wife of a Mongolian general. Soon, she was able to perform archery and martial arts with her horse at a gallop as deftly as she was able to compose poetry or analyse the classics of Chinese literature, and by 16 she was ready to accompany her father on trips all across China, absorbing what she saw of local economic and social conditions, and employing her observations of injustice and suffering to deepen her poetic output.

Wang Zhenyi left behind thirteen volumes of poetry, which deserve their own analysis altogether, but today we are going to focus on her contributions to mathematics and astronomy, and to feel the impact of those, we need to ask the question, just where was China scientifically by the mid- to late eighteenth century? I think we have a tendency, particularly in the West, to think of Chinese science as consisting of an explosion around the Tang (618–907) and Song (960–1279) Dynasties, then a steady decline into medieval stasis that was not reversed in any substantial way until the tail end of the Qing Empire. In reality, by the eighteenth century, the Chinese astronomical revolution was well underway, supported by three successive waves of imported experts who brought with them the most advanced insights of their scientific traditions.

Once Buddhism started making its way into China across the Silk Road in the first century CE, carried by Indian missionaries, it was not long until other aspects of Indian culture began making inroads into Chinese intellectual life as well, and principal among these was India's pre-eminence in mathematics and astronomy. This tradition reached its culmination in the work of Aryabhata (476–550) whose development of trigonometry and application of its principles to spherical trigonometry allowed a powerful mathematisation of the night sky a millennium before the efforts of Galileo and Newton.

The Chinese imperial government took note of the accuracy of Indian astronomical techniques and in the seventh century a number of Indian astronomers could be found working in the provincial capital of Chang'an. Their work was then amplified by a wave of Islamic astronomers brought in by the Yuan/Mongol Dynasty in the thirteenth century as part of their racial policy of staffing the upper echelons of the Chinese bureaucracy with as many non-Chinese individuals as they could get their hands on. Persia was one of their particularly preferred hunting grounds for imperial administrators and officials, and from Persia came Jamal ad-Din in 1267, carting with him a terrestrial globe, an armillary sphere, and an Islamic astronomical almanac. His command of astronomical technology, and success as the director of Beijing's Islamic observatory, impressed the Yuan, who took to importing more Islamic scientists and inventing their own variations of Islamic astronomical instruments, a practice which continued to the Ming Dynasty (1368–1644), with the erection of more observatories and compilation of new observation charts and calendars.

The ground was set, then, for the third wave of foreign inspiration, as European astronomy made its way to China by way of the Jesuits in the sixteenth century. They brought with them the telescope, and the complicated mission of introducing China to the great advances being made in post-Copernican European astronomy without, however, betraying the Catholic Church's strict adherence to a geocentric model of the universe. Eventually, some Jesuits broke ranks and shared the work of Copernicus, Galileo and Brahe with their Chinese colleagues, creating new tensions in Chinese scientific life. Some, like Xu Guangqi (1562–1633), not only eagerly advocated the new Western ideas but converted to Catholicism as well. Some, like Mei Wending (1633–1721), advocated for a fusion of Chinese and Western practices that recognised and utilised the best of both traditions, while others, like the Yongzheng Emperor (1678–1735), who proscribed Catholicism outright in 1724, thought the whole friendship with the Jesuits and their Western innovations was a mistake that needed to be quashed if China was to return to its habitual stability.

This was the intellectual climate into which Wang Zhenyi strode as a woman in her early twenties, possessing depths of both life experience and scholarly learning that few could claim even in their sixties and seventies. She studied Mei Wending's *Principles of Calculation* and distilled its lessons into a more approachable format she called the *Musts of Calculation*. She also wrote *Supplementary Information on Western Calculations*, which demonstrated her firm advocacy of any body of knowledge which worked, regardless of the ideology of those who created it, as well as the five-volume *Simple Principles of Calculation* and four-volume *Beyond the Study of Mathematics*. In this same spirit, she argued for the adoption of Western calendrical systems, stating, 'What counts is the usefulness, no matter whether it is Chinese or Western.'

Similarly, when approaching the cosmos, Wang Zhenyi did not chain herself to any intellectual ideology but seems to have followed where her reasoning and study led. She described lunar and solar eclipses from a heliocentric perspective, and built a model of the Earth-Sun-Moon system in her garden pavilion as a way of testing and demonstrating how eclipses worked under Copernican assumptions. She also argued for a spherical Earth, and took pains, without the benefit of the Newtonian theory of gravity (the *Principia* was not available in a Chinese edition until the 1850s), to explain how the world's population could live on a globe without half of them falling off. Even more than her work in mathematics, her work in astronomy contained not only helpful and accessible summaries of past research but bold new ways of thinking about the cosmos from the perspective of mathematical analysis that answered existing mysteries while posing fruitful new questions. In addition to her articles on equinoxes, eclipses and the shape of the Earth, she wrote a larger text, *The Explanations of Constellations*.

This was a remarkable body of work for a person not yet 30 years old, and it is one of the great tantalising questions in the study of science as to what she might have accomplished given a longer life; would her knack for mathematics and astronomy, and her curiosity about the workings of what we now call gravity have made her the Newton of the East in her time? We will never know, and truthfully, we are lucky we know as much as we do about her, for as she was dying she entrusted her life's work to her best friend, who, in turn, passed on those manuscripts to the scholar Qian Yiji, who arranged for the publication of the *Simple Principles of Calculation* and preserved (most of) the rest to ultimately land in the private collection of Zhu Xuzeng. With so many priceless documents being transferred between so many hands over so many centuries, we should rest grateful for having as much of her output as we do, with its daring breadth of scope and refusal to bow before any manner of convention, be it social or academic. She trusted the evidence of her eyes and the guidance of her reason, and became one of the greatest scholars and writers of her era on the strength of them. In the words of one of her own poems, which also ranks as a mic drop of world-historical proportions:

> Are you not convinced
> Daughters can also be heroic?

Further Reading

In English, there is pretty much just one book that every article about Wang Zhenyi pulls from, and that is the account of her in Barbara Bennett Peterson's *Notable Women of China: Shang Dynasty to the Early Twentieth Century*. It is pricey, and if you are looking solely for women in science, she's the only one in it, but the other portraits, of scholars and poets and performers and warriors, are fascinating as well. If you can read Chinese then Shen Yu Wu's *Female Scientists in the Qing Dynasty* is likely the book for you. Currently, it looks down from the shelf at me with accusation, but someday perhaps two decades hence I will scratch together enough fluency in Chinese to make a start on it, then I will tell you how it is!

Chapter 7

Brief Portraits
The Early Modern Era

Sophie Brahe (1556–1643)

Sophie Brahe was not unique for her time in the widely disparate experience of heady intellectual highs and miserable personal lows. After all, the last years of her life map almost entirely upon the great demographic implosion of the Thirty Years War (1618–1648) which devastated vast swaths of Europe, and laid low more than a few once noble families. And yet, while her life was perhaps not uniquely tragic in a material sense, few women of her age probably experienced quite the magnitude of intellectual fall of Brahe, starting from such an elevated position, with a Europe-wide reputation for scientific achievement in multiple fields, before plummeting to the depths of scorn in the role of spouse to a failed and bankrupt alchemical speculator, and ultimately achieving some level of equilibrium only after gaining her independence in later life.

Brahe was the daughter of Beate Bille Brahe, who managed the household for Queen Sophie of Denmark (herself one of the most learned women of her time, with a deep interest in astronomy), and Otto Brahe, an advisor to King Frederick II of Denmark, and governor of Helsingborg castle. She was born either in 1556 or 1559, making her either 15 or 12 when her father passed away, which allowed her to study more freely whatever subjects interested her, and also brought her under the educational encouragement and guidance of her older brother, Tycho Brahe.

Tycho noticed her genius, and supported her study of German, Latin, chemistry and literature, though he at first was resistant to her following his interest in astronomy, feeling that it was too intellectually challenging a topic for a woman to grasp. He was soon disabused of this notion, however, and over the course of the 1570s Sophie was a regular assistant in his laboratory, including during the crucial observations of the 1572 supernova which were detailed later in Tycho's *De nova stella*. That supernova, one of only eight that have been recorded in history as visible to the naked eye, sent shockwaves through the astronomical community, presenting as it did a new resident in the theoretically eternal and changeless night sky of Christian theology. The geocentric model, it seemed, had to go, but Tycho and Sophie were not entirely convinced of the correctness of the Copernican system that sought to replace it. They developed their own compromise solution, the Tychonic Model, which featured the Sun and Moon orbiting the Earth, but the other planets orbiting the Sun. Some have hypothesised that Sophie might have had a significant part to play in the development of the model, as Tycho's strengths lay in the fields of instrumentation and observational management, while she was the stronger of the pair mathematically, but really there is no way of knowing for sure.

In 1579, Sophie married a man named Otto Thott, who was either ten or thirteen years her senior, and who died in 1588. After his death, Sophie deepened her medical and chemical studies, managed Tycho's observatory and research programmes when he was compelled to travel, and gained a wide reputation for garden design, while also visiting her brother regularly at his observatory. Here she met Tycho's friend Erik Lange, a nobleman with a passion bordering on mania for alchemy. Sophie and Erik became engaged in 1590, but because he had driven himself to poverty funding his alchemical pursuits, he spent the next decade largely scuttling through northern Europe attempting to avoid his creditors, and the couple did not marry until 1602. It was not a particularly joyous affair – the Brahe family distinctly disapproved of Erik, and refused Sophie any further financial support, compelling the couple to get married in rather tattered and borrowed clothes.

Eleven years later, Erik was dead, and Sophie could begin the process of rebuilding her finances and reputation. It was during this time that she compiled a 900-page genealogical account of Scandinavian royal families, published in 1626 and still a consulted source for early modern Danish history. She also continued various scientific researches, particularly in medicine, to the benefit of her local community. She lived long enough to see the data compiled by her and her brother over so many decades put to use by Johannes Kepler to construct mathematically rigorous models of the solar system that fundamentally and forever changed Europe's conception of the cosmos, and humanity's place within it.

Elisabetha Catherina Koopman Hevelius/Elzbieta Heweliusz (1647–1693)

When Elisabetha Hevelius published *Prodromus astronomiae* in 1690, she was not only providing the world with a definitive catalogue of over

1,500 stars, some 600 of which had not been until that time calculated and published, but she was also honouring the project that she and her husband had set out on some three decades previously, and which he had died before seeing to completion. Hevelius was the daughter of a successful merchant family in Danzig, which was at the time one of the most prosperous and largest commercial cities in the Hanseatic League, a collection of northern European territories participating in free trade and mutual protection.

As a child, Elisabetha visited the sprawling observatory of Danzig's foremost astronomical scientist, Johannes Hevelius (1611–1687). It spread across the connected rooftops of three adjacent buildings, and struck the young Elisabetha as one of the most marvellous things she had ever seen. Johannes, for his part, noted her enthusiasm, and promised to teach her about astronomy when she was older. How that education proceeded we do not know, but in 1662 Johannes' wife died, and the next year he married Elisabetha. She was 16 and he was 52, which is often a recipe for misery for all concerned, but by all accounts worked out well enough in their case. When Elisabetha contracted smallpox, it was Johannes who stayed by her side and nursed her back to health, and there is nothing in the literature which suggests anything but happiness through a joint devotion to the exhaustive study of the night sky (though, of course, the nature of what got recorded and what did not in the seventeenth century was such that positive accounts of the relationship were far more likely to get transmitted to posterity than negative ones, so we should always be a little careful with taking those sources at their word).

The couple's main project during their years together was a massive naked-eye survey of the night sky, with the object of not only finding and recording new stars but verifying the positions of those already recorded in the great Brahe and Ptolemaic data sets from the past. Elisabetha made observations and calculations, and managed the running of the observatory. The whole project nearly came to a disastrous end in 1679, when a fire raged through the observatory, but fortunately a notebook containing the couple's most important computations and positions survived the blaze, and a series of royal patrons (including Louis XIV and John III of Poland) provided funding to get the observatory quickly back on its feet. The couple continued their work for the next eight years, rebuilding lost data and pressing forward with new observations. Even though they possessed Europe's most powerful telescope, they declined using it for this particular study, trusting to Johannes' almost legendarily sharp eyesight over the possible errors that could be introduced to observation by imperfections in the age's lens technology. Unfortunately, Johannes died before the work was complete, and it was left to Elisabetha to finish their life's work, publishing the *Prodromus* in 1690, which included not only tables listing 1,564 stars, as recorded by the Heveliuses, and comparisons with the positions computed by earlier astronomers, but also a fifty-four-page atlas detailing seventy-three constellations of the Northern and Southern hemispheres.

Jeanne Dumée (1660–1706)

Jeanne Dumée is a figure unlike any other in the early modern era of astronomy. Whereas the post-Cunitz figures we have visited so far, and those we shall meet over the course of the next century and a half, were, generally speaking, attached to the observational efforts of some male astronomer as an assistant, engraver, or computer, Jeanne Dumée equipped and ran her own observatory. She was able to do this because her husband, a military officer of apparently some means, died when she was 17, leaving her with sufficient resources to not only subsist without the financial necessity of remarrying but in fact to fund her own scientific endeavours. She visited the roughly two dozen private observatories in Paris that existed at the time, including a tour of the Royal Observatory, which had been established in 1667 and was up and running by the early 1670s.

Dumée was a devoted Copernican in a time when the debate between the Ptolemians and the heliocentrists was still very much alive (Galileo died just eighteen years before Dumée was born, and the works of Kepler in defence of Copernicus were a mere half a century old). With characteristic aplomb, Dumée strode straight into the centre of this debate, hosting salons and public lectures, and building her own armillary spheres along the Copernican and Ptolemaic lines to demonstrate the superiority of the heliocentric system. This was the work of a public scientific intellectual, a role not open to most women of her time, and she followed it up with at least one publication, *Entretien sur l'opinion touchant la mobilité de la terre* (1680), a pro-Copernican treatise which we have in manuscript form, but no surviving actual copies of.

Margaret Flamsteed (*c.* 1670–1730)

Margaret Flamsteed was the wife of John Flamsteed (1646–1719), the first man to hold the title of Astronomer Royal in Britain and a guiding force behind the establishment of the Royal Greenwich Observatory (which did not have a woman director until 1972, when Margaret Burbridge was tapped for the job) in 1675. We know little about her other than that she was well educated, and played a key role in ensuring that two of John's most important books, *Historia Coelestis Britannica* (1725) and the *Atlas Coelestis* (1729) were published after his death, a costly process that all but drained the finances left her after John's death in 1719.

Maddalena Manfredi (1673–1744) and Teresa Manfredi (1679–1767)

There are two famous pairs of astronomical sisters in this book, and while the duo we shall meet later, the Kirch sisters, are known to us as individuals with their unique interests and life trajectories, the Manfredi sisters come to us as an irreducible unit, differing in birth and death years, but otherwise indistinguishable in the historical record. Though we know a good deal about the impressive literary and scientific output of the pair, we have no way of saying which among them performed what part of their joint labours. But that's okay, if there is one thing that I have learned raising two daughters, it is that they *love* sharing credit for work.

Maddalena and Teresa were two of the seven remarkable children born to Alfonso Manfredi and Anna Maria Fiorini. Alfonso was a notary, which meant he had the resources to send some, but not all, of his children, to institutions of higher education, and that they could not expect endless parental resources to follow whatever individual dreams they might have. This, in combination with the unusual degree of genius among them, seems to have pushed the Manfredi siblings together as a cohesive unit. Only one of the seven ever married, while the rest (save for one rogue Jesuit among them, the underwhelming youngest sibling Emilio) pooled their resources and talents in the pursuit of various scientific goals. Gabriele was an important figure in the popularisation of differential calculus in its Leibniz-Bernoulli form. Eustachio, the eldest, was a key figure in the early history of the Academy of Bologna and an astronomer. Meanwhile, Eraclito (sometimes given as Heraclitus) studied medicine before ultimately turning to physics and astronomy.

It was a talented bunch, with Maddalena and Teresa sticking particularly close to Eustachio, who found a powerful patron in the form of Luigi Ferdinando Marsili, a soldier and scholar who saw service in the Austrian wars against the Turks, and employed his scientific skills both as a military engineer and in the sensitive process of determining the Ottoman-Habsburg border after the Ottoman's final defeat at the Battle of Zenta in 1697. Marsili's goal was to make his native Bologna a scientific capital to rival Paris and London, but unfortunately at the time the University of Bologna was in a particular lull intellectually, and so Marsili decided to reach out to people outside the institution who were scientifically inclined. Eustachio Manfredi had, at the age of 16, organised a scientific circle, the Academy degli Inquieti, in 1690, which had become an academic and scientific salon of sorts. Marsili selected Eustachio to establish his prospective Bolognian Academy, and in either 1700 or 1704 the Manfredi sisters followed him as he set up shop in Marsili's palace.

Eustachio was made head of the Academy's Observatory (which resided in the third floor of the institute's home in the Palazzo Poggi) in 1711, and held that position until his death in 1739, during which span he relied, as he always had, on his sisters for their aid in observation and computation. They performed longitude and latitude calculations along with the computations for planetary trajectories and stellar positions that underpinned Eustachio's *Ephemerides motuum Coelestium*, published from 1715 to 1725 and which ranked among the most trusted sources in Europe for determining terrestrial position from celestial observations, as well as for accurate astronomical observation. In addition to their work on this and subsequent astronomical publications for their brother, the Manfredi sisters also made a name for themselves as translators of Neapolitan stories, publishing several popular editions in the early 1740s. We do not have any record of Teresa's work after Maddalena's death in 1744, and there's no way to interpret that fact that comes to mind that isn't quite sad indeed.

Marie Claire Eimmart (1676–1707)

Somewhere in the Museum of the Observatory in Bologna, there lie ten pieces of scientific art, exquisite relics on dark blue cardboard of an age before photographic plates, when an astronomer, in order to fully share her observations, needed as much of an artistic hand as she did a discerning eye. They represent a small fraction of the output of Marie Claire Eimmart, who was heading towards a reputation as one of Europe's great astronomical observers and engravers when her career was cut short at the age of 31 in childbirth. Both her grandfather, Georg Christoph Eimmart the Elder, and her father, Georg Christoph Eimmart the Younger (1638–1705), were famous artists and engravers based out of the Bavarian city of Nuremberg. Marie Claire's father was a passionate amateur astronomer and sunk the profits from his artistic ventures (including his salary as director of the Nuremberg Academy of Art) into the purchase of astronomical instruments, which formed the basis of the private observatory that he built for himself in 1678, and in which Marie Claire worked as an assistant.

Georg Christoph taught Marie Claire mathematics, Latin, French, astronomy, drawing and engraving, and during their years together she produced hundreds if not thousands of representations of terrestrial and astronomical phenomena, including over 300 drawings devoted to the minute rendering of the Moon's phases. In 1705, Georg Christoph died and shortly thereafter the city of Nuremberg purchased his observatory, and named his student, Johann Heinrich Muller, as the new director. The following year, in 1706, Muller and Eimmart married, though the couple would have but little time together. In 1707, Eimmart and her baby died in childbirth.

Of the hundreds of drawings produced by Eimmart during her short life, only a handful survive today, the bulk being the ten drawings which Georg Christoph had sent to his friend, Luigi Marsili (the same scientific patron we met in the story of the Manfredi sisters), who subsequently donated them to the observatory of Bologna, and which depict phases of the Moon and Mercury, aspects of Mars and Saturn, and various representations of comets, all of which speak to a more personal and direct era of astronomy, when the cosmos spoke to the eyes, and the eyes guided the hand, in a perfect fusion of observation and art.

Christine Kirch (1696–1782) & Margaretha Kirch (1703-sometime after 1744)

The two daughters of Maria Winkelmann-Kirch led lives that began from very similar positions but reached quite different ends, to the point that one of them was, by the end of her life, celebrated in her time by the academic community for her scientific contributions (and in ours for her position as the world's first paid woman astronomer), while the other has steadily faded from our historical memories.

Let's start with the forgotten sister, Margaretha. She was only 7 years old when her father, the astronomer Gottfried Kirch, died, leaving her mother to support the family as best she could within the limits of the Berlin Observatory's willingness to give her a position and salary commensurate with the work she was doing. Both Margaretha and her sister Christine were denied the formal education granted as a matter of course to their brother,

Christfried, and were instead trained by their mother from the age of 10 to participate in the family business of astronomy. When their father died, and their mother was prevented from assuming his position, the family's fate eventually fell to their brother, who took over his father's old position at the Berlin Observatory in 1716. Margaretha, for her part, in addition to the usual computation duties to be expected of her in service to her brother's work, seems also to have been engaged in meteorological observations, as well as studies of the night sky that allowed her, on 3 January 1744, to view the Great Comet C/1743 X1, which she continued tracking over the following months. Though she made important sketches and observations of the comet, including the phenomenon of its split tail, she was not, as you will often read online, the discoverer of that comet, as it was first observed by Jan de Munck in late November 1743.

That is the sum total of what we remember now of Margaretha – whether she continued her work for decades like her sister, or died in April of 1744, we simply do not know.

That brings us, then, to Christine, who as the older sibling had more time with their father than Margaretha did, and whose life has been much more thoroughly documented, as she spent it doing recognised and lauded work for major astronomical institutions. As far as we know, she had the same training as a child as Margaretha did, but when it came time to aid her brother at the Berlin Observatory, it seems that she had a much larger role in the annual preparation of the Berlin Academy of Science's calendar (which was one of the Academy's biggest money-making ventures) than did Margaretha, and was therefore that much more connected to people in positions of scientific authority than her sister. When Christfried died in 1740, Christine became even more important to the Academy as the only person with the experience and knowledge to keep the calendars running on time, an importance which increased with Frederick the Great's invasion of Silesia in 1740. His eventual annexation of that territory, and with it a significant Catholic population, required a slightly different calendar, which it fell to Christine to produce.

For the next three and a half decades, Christine worked for the Academy as the reliable scientific labourer behind one of their most important regular projects, all without a salary. She was instead voted occasional 'donations' from the Academy, a situation which became increasingly absurd until finally the Academy rectified it in 1776, when Kirch was 80 years old and had been working for them for virtually her entire life, by voting her an annual salary of 400 thalers. Assuming these were Reichsthalers, which had been adopted in 1750 and which contained 16.70 g of silver, which sold for roughly 5 shillings to the ounce at the time, this would have represented a yearly salary of some 59 pounds, or approximately 5000 pounds in today's currency, which was a meager amount for the work that she did, but a historic one.

Whatever she was actually paid, the important point here is that Christine *was* paid a regular salary, and became thereby the first woman in modern history to earn a salary through her astronomical work. Eventually, she was able to collect this salary even without the obligation to continue churning out calendars which, really, was the only decent thing for the Academy to do. There is an inconsistency, however, in Christine's story here, as most sources state that she received her Emeritus status from the Academy at the age of 77, and with it the ability to draw her salary without producing the calendar, but that would have been in 1773, i.e. three years *before* she is recorded to have *started* receiving her salary which appears impossible without the aid of time travel.

Nicole-Reine Hortense Lepaute (1723–1788)

Nicole-Reine Lepaute was, with Madame du Chatelet (whom we shall meet in the Mathematics volume of this series), France's great example during the High Enlightenment of what women could achieve in science if offered the slightest encouragement and guidance. Her family were servants of the royal court in the years of the regency preceding the formal ascension of Louis XV. Renowned for her intellect and studiousness, she was not long in finding a similarly driven employee of the Court, the royal clockmaker Jean-André Lepaute, whom she married in 1748, and convinced to turn his talents towards the design of an astronomically oriented clock of her conception. That clock was presented to the French Academy of Science in 1753 and caught the eye of Joseph-Jérôme Lalande, an up-and-coming astronomer in his early twenties who would go on to be one of France's most important scientists, and who quickly befriended the clock-making couple.

Lepaute began her scientific investigations with an exhaustive study, calculating the length of a pendulum required to produce every oscillatory frequency between 1 and 18,000 cycles per hour, at intervals of 100 oscillations. This involved 180 sets of hand computations, and in 1755 she published her results as the *Table des longueurs des pendules*. At this point, she and her husband moved more closely into the orbit of Lalande's interests. It was only natural that an astronomer would befriend one of

Europe's most respected clockmakers – a good clock was perhaps second only to a good telescope in terms of necessary gear for the astronomical trade. What was truly special, however, was Lalande's complete and instant acceptance of women as men's potential scientific equals. Later in life he stated, 'I believe that what women are missing is merely educational opportunities and examples to emulate. We see them sufficiently successful, despite obstacles of education and prejudice, to believe that they have just as much talent as most men who have gained a scientific reputation.' Lalande was central to the encouragement and employment of three of France's greatest women astronomers in the late eighteenth century – Lepaute, Louise du Pierry, and Marie-Jeanne Lalande, the latter two of whom we shall meet in due course.

In 1757, Lalande decided to undertake the seemingly mad project of calculating the date of Halley's Comet's return, factoring in all of the accumulated gravitational influences from Jupiter and Saturn over the preceding seven decades. This involved computing, moment by moment, the combined gravitational force on the comet from the Sun and any relevant nearby planets significantly tugging on it, to find the combined net force on the comet, and determine the deviations from a perfectly elliptical path caused by that net force. Lalande brought Lepaute and the mathematician Alexis Clairaut into the plan, and together the bold trio crunched numbers during every waking hour over the course of six long months to get a result before the comet's expected return. They just managed the task, a feat that Lalande said would not have been possible without Lepaute, and were able to announce to the Academy of Sciences that the old accepted return date of 1758 calculated by Edmond Halley was in error, and that the actual perihelion would fall somewhere between mid-March and mid-May 1759. When the comet returned to its perihelion position on 12 March 1759, it was a triumph not only for Lalande, Clairaut and Lepaute but the entire venture of predictive astronomy for perturbed orbits.

In 1762, Lepaute solidified her reputation for rigorous predictive computation with the publication of the snappily titled *Explication de la carte qui représente le passage de l'ombre de la lune au travers de l'Europe dans l'éclipse du soleil centrale et annulaire de 1 Avril 1764*. This book undertook to predict the duration, magnitude and path of the upcoming 1764 solar eclipse, working out how it would appear and progress, not only to observers in Paris but to those in other major European countries as well.

Lepaute was also, for a period of some two decades, the primary calculator for France's most important astronomical publication, the *Connaissance des Temps*, which had been established in 1679 to annually publish the positions and trajectories of the stars and other astronomical objects for navigational and astronomical purposes, as well as to present new results in astronomical research. (It is still published, by the by, under the title *Ephemerides astronomiques*, making it the world's oldest and longest running publication of ephemiredic data.) Lalande edited the *Connaissance* from 1760 to 1775, and Lepaute worked as the primary calculator for this publication from 1759 to 1774, when she switched over to doing similar computations for the Academy's publication *Ephémérides*, continuing that work until 1783.

In 1767, her husband began showing signs of a terminal illness, and the remaining twenty-one years of Lepaute's life were increasingly devoted to caring for his health needs. This service, combined with the demands of her astronomical efforts, took a physical toll, and in late 1788 Nicole-Reine Lepaute lost her eyesight, and then in December of that year, her life. Jean-André followed her three months later.

Louise Elizabeth du Pierry (1746-*c*. 1807)

Middle children tend to get overlooked – while the first gets lodged in memory for all the anxiety and joy that comes with a brand-new experience, and the youngest represents a bittersweet last chance to get things right, the ones in between are, well, easy to lose track of. Likewise with the historic grand trio of French women astronomers who between them drove their country's astronomical computations over the course of the entire late eighteenth century, and into the early nineteenth. We remember Nicole-Reine Lepaute, the royal servant's daughter turned scientific trailblazer, and Marie-Jeanne Harlay, the calculational powerhouse who closed out the era, but tend to forget Louise du Pierry, whose life path wove itself neatly between theirs and whose work was, if anything, the most broad in scope of the three.

We do not know a great deal about her life, and some of the *facts* offered in online accounts of du Pierry are more educated inferences than solid data. We know she was born in Maine province, married at age 20, and met Lalande in 1779. Some authors, pointing towards Simone Dumont's 2007 biography of Lalande, assert that du Pierry was Lalande's mistress, but I think there is room for some scepticism with that particular claim. Another source of biographical confusion is the status of a women's course in astronomy she offered in 1789, which some authors assert was officially offered through the Sorbonne University. Given that the Sorbonne did not promote a woman to a professorial position in science until 1906, and only then when it was Marie Curie herself assuming the position of her recently dead husband, that would have been quite the academic event. More likely, unfortunately, is the explanation that the class was held at du Pierry's own initiative, within her private rooms.

When we come to du Pierry's scientific accomplishments, however, we return to stable ground. She worked as a computer and compiler for both Joseph-Jérôme Lalande and the chemist Antoine Fourcroy, doing the long hard work to produce the tables of scientific values that formed the indispensable practical baseline of a scientist's work. For Lalande, her first publication involved exhaustive computation of the refractive effects caused by the Earth's atmosphere, which shifts and distorts the apparent positions of celestial objects away from their true locations. In her 1791 *Tables de l'effet des réfractions,* du Pierry worked out how bad the refractive effect would be based upon where one was, the atmospheric conditions prevailing there, and in what direction one was pointing one's telescope (the more a telescope deviates from looking straight upwards, the more atmosphere it has to look through to see the night sky, and therefore the more pronounced the refractive effects).

The refractive tables were a grand service to astronomers everywhere, and du Pierry was not nearly done. In her 1792 *Tables de la durée du jour et de la nuit,* she calculated the expected day and night lengths for the near future, which were valuable not only to astronomers, but to civic officials as well, which was a wise move considering the fate that befell some scientists during the French Revolution, and particularly the Terror, who could not prove the practical merit of their work for the people. Whether out of political cunning or honest curiosity, du Pierry continued this trend of work that had one foot in the scientific and the other in the civic world with her third book, *Calculus d'éclipses pour mieux trouver le mouvement de la Lune.* The orbit of the Moon is an important thing to know accurately, as from it flows knowledge of the tides, which are in turn a key component in the safety of maritime operations – scientific, military and merchant. Du Pierry's strategy of using rigorous computations based on centuries of accumulated eclipse data as a gateway to deciphering lunar motions more finely was an intriguing direction from which to tackle the problem, one which made particular use of her computational gifts.

Whereas most of Lalande's calculators stuck closely to his work and research interests, or those of approved astronomy-related colleagues, du Pierry, in her fourth major work, left the astronomical nest to pursue work in one of France's most rapidly developing fields in the late eighteenth century – chemistry. The father of modern chemistry, Antoine de Lavoisier, was guillotined in 1794 for tax fraud, but his student Antoine-Francoise Fourcroy not only survived (some say by throwing his teacher and colleague to the wolves while doing as little possible to help him) but rose to the level of State Chancellor and eventually, on the day of his death in 1809, to the status of Count. He is known today for his efforts to standardise chemical nomenclature after centuries of alchemical chaos, and to improve medical practice through more chemical research into the properties of different biological tissues. His last major published writing was a ten-volume behemoth released in 1801, the *System of Chemical Knowledge*, for which du Pierry compiled and wrote a 170-page table that organised and systematised all of the different materials spoken of in that ten-volume work. This was her last major publication, and after that, the woman who had made such an impact on French science that Lalande dedicated his 1785 celebration of women astronomers, *Astronomie des dames*, to her, faded away so precipitously that we cannot even say for certain the manner, place, or year of her death.

Mary Anne Fallows (née Hervey) (*c.* 1750–1838)

When it was finally completed in the late 1820s, the Royal Observatory at the Cape of Good Hope was the first permanent observatory to be established in the Southern Hemisphere, and the brunt for pulling off that historic first was entirely borne by two stalwart scientific devotees: Fearon and Mary Anne Fallows. They arrived at the Cape in 1820 and found nobody there to help them unload their precious cargo or show them where to go, a trend in neglect that would continue over the seven long years of getting the observatory up and running in the face of financial neglect from England, supply chain problems, and indifference from the local government. Observations at the observatory required two people to carry them out, and when in 1830 Fearon's assistant left the dreary environs of the Cape, it was up to Mary Anne to fill in, not only taking up the assistant's role but also parts of Fearon's as his health steadily deteriorated, culminating in his early death in 1831 at the age of 43. It fell to Mary Anne, then, to gather what observations they had made, and return to England to oversee their preservation. Having ensured that the proper computations were done to reduce the raw Cape data from 1829–1830 to usable positional information, she more or less left the field of science, except to advocate for the official publication of the Cape observations. She remarried in 1835, and passed away three years later from a freak self-prescribed leech regimen accident, as was the custom of the time.

Margaret Bryan (*c.* 1760–1816):

In many ways, Margaret Bryan's story has faded from view largely because her life and work overlapped with those of a few other key women scientists in precisely a manner so as to appear almost invisible. She was an English woman astronomer active during the era of Caroline Herschel, whose possession of the Herschel name and access to the resources that came with it assured a measure of fame that was not approachable for most women. She was also a gifted English scientific educator whose major works, however, came just a decade or two before those of Mary Somerville, whose titanic presence as a science communicator tended to overshadow just about anybody in her intellectual vicinity. Pushed back by the fame of Herschel and Somerville, the area that Bryan *could* call uniquely her own was her work in advocating for the scientific education of girls, and her role in establishing three separate schools for girls (at Blackheath, Hyde Park and Margate), but of course scientific historians until very recently have had such a prejudice in favour of scientific theorists over important scientific educators that her pedagogical inventiveness was summed up in a historical shrug for two centuries.

Let's do a bit better by Margaret Bryan, then. She published three textbooks that aimed to bring recent developments in astronomy within the grasp of students and enthusiastic lay people by teaching their important ideas visually (often in her own drawings), rather than purely mathematically. In *A Compendious System of Astronomy* (1797), Bryan talked about Newton's optics and laws of motion, eclipses, the Copernican system, fixed stars, the discovery of Uranus in 1780, gravity, and lunar motion, all in an engaging conversational style. She followed that volume in 1806 with *Lectures on Natural Philosophy* which discussed a wide array of topics in physics and astronomy, including fluids, magnetism, electricity, acoustics, pneumatics, pressure and optics. Finally, in 1815, she published the *Astronomical and Geographical Class Book for Scholars*. Perhaps my favourite of her efforts, however, to bring astronomy to everyday people was a board game that she had a hand in, 'Science in

Sport or the Pleasures of Astronomy'. Originally issued in 1804, Bryan played some editorial role in the 1815 reissue, in which players advance along the board, often by reciting scientific facts, like the properties of the telescope or the phases of the Moon. Landing on certain tiles caused players to get stuck (as in the 'county gaol' space reserved for people who spent more time calculating the trajectory of billiard balls than those of planets) or instantly lose the game (a square depicting a 'blockhead' boy which is perhaps my favourite).

Bryan was an educational innovator, finding new ways to reach students with the wonders of science, some of which have taken us a good couple of centuries to catch up with, and for the sake of the 150 students she taught directly, and the untold number more she taught through her words, it is probably time we find a place for her in the regular pantheon of English scientific legends.

Marie-Jeanne-Amélie Harlay de Lalande (1768–1832)

The familial background of Marie-Jeanne Harlay de Lalande is an anarchic fun-house almost tailor-made to drive historians mad. No two sources say quite the same thing about the relations between the key characters in her life. For many years, the closest you could get to a standard story was that she was the illegitimate daughter of Joseph-Jérôme de Lalande (whom we met above), whom he called his niece to avoid the shame of illegitimacy, and that she married Michel Lalande, who was Joseph-Jérôme's cousin, but who Joseph-Jérôme called his nephew, presumably so that Michel and Marie-Jeanne would thereby be second cousins, rather than first cousins once removed. In 2015, however, new research revealed that this complicated story was probably nowhere near the truth, and that Harlay was in fact the daughter of Jean Francois Harlay and Anne Elizabeth Cany, and that her marriage to Michel was therefore entirely lacking in salacious scandal.

I know, rather disappointing, really! What is not at all disappointing, however, is the quality of her astronomical work, which consisted of a number of truly monumental feats of computation. The first, published in 1793, was a 300-page collection of tables for Lalande's *l'Abrégé de navigation*. This work allowed sailors to calculate their position on the globe with reference to the elevation of the Sun and certain key stars, and was so not only an impressive display of astronomical skill but one of crucial practical importance in an age of increasing intercontinental travel.

More epic still in scope were Harlay's efforts in the mid- to late 1790s, performing the computations that reduced Joseph-Jérôme and Michel's observations into reliable star positions, and which by some estimations were carried out by Harlay at the rate of some 12,000 per year during the height of their work together, with each of those reductions requiring some thirty-six separate computations, all done, let us remember, entirely by hand. As a result of this work, Harlay was able to independently publish a collection of 10,000 star positions she had computed in 1799, and hers was the main force behind the crown jewel of Lalande's late career, the *Histoire Celeste Francaise* of 1801 which featured the positions and magnitudes of some 47,000 stars. As if that was not enough, from 1794 to 1806 Harlay was *also* responsible for the annual calculation and publication of the *Connaissance des Temps*, which featured updated information on the paths of the most prominent objects in the solar system.

And as if all of *that* was not enough, Harlay also gave birth to three children, a daughter born on the anniversary of Caroline Herschel's first discovery of a comet, who was named Caroline in her honour, a son Isaac (named after Isaac Newton), and a second daughter who died in childhood named Uranie, named after the Muse of Astronomy who also features in the title of Maria Cunitz's *Urania Propitia*.

Maria Short (*c.* 1788–1869)

In 1776, as the American colonies were declaring their independence from Great Britain, a man on a hill in Edinburgh was trying a somewhat revolutionary experiment of his own. His name was Thomas Short, and he was widely reckoned one of the finest makers of scientific equipment in all of Britain. He had the idea that the public might be willing to pay to have access to astronomical equipment, and in 1776 he leased some land on Calton Hill with the intent of erecting a commercial observatory there which would feature some of his most renowned instruments, including what was simply called the Great Telescope. One of the conditions of his lease, however, was that the observatory, and everything in it, could not be inherited by a female descendant.

Twelve years later, in 1788, Short died, his pet project still unfinished. Though his grandson carried on with the work, and opened the observatory in 1792, it closed shortly thereafter, not being the profound commercial success Thomas had hoped for. In the ensuing decades, Edinburgh decided to build its own observatory, which is still in operation, on a nearby site, and that seemed to settle the whole matter. That is, at least, until 1827 when a woman arrived in Edinburgh from the West Indies claiming to be Maria Short, the daughter of Thomas. She claimed ownership of the Great Telescope, and returned to Calton Hill, debuting Short's Popular Observatory in 1835. For a small fee, the general public could enter and learn about the planets and experience a camera obscura show that depicted the Edinburgh skyline. The forerunner of the modern private planetarium, the Popular Observatory was a hit with the public for fifteen years, drawing thousands of visitors until the Edinburgh local council ordered it pulled down in 1851.

Eléonore-Nell-Suzanne de Bréauté (1794–1855)

De Bréauté was a member of the French nobility, the daughter of Jean Susanne de Bréauté (1772–1847) and Marie Françoise Eléonore Josèphe Letellier d'Irville (1756–1818). We know very little about her astronomical life, save for the fact that she was the first to witness the Great Comet of 1823, her observation of 29 December preceding those of Jean-Louis Pons and Wilhelm von Biela, who both witnessed it on 30 December. She also constructed a topological map of Normandy, and financed Jules de Blosseville's ill-fated 1833 expedition to map the east coast of Greenland. That expedition disappeared in August 1833, never to be found again, though its memory lives on in the name of Greenland's Blosseville Coast. De Bréauté, for her part, also earned a bit of geographical immortality in 1823 when the Bréauté Peninsula in Normandy was named in honour of her cartographical work. In 1888, she was the subject of a twenty-four-page biographical sketch by Jean Benoît Désiré Cochet which forms the basis for more or less all we know about her today.

Chapter 8

Computing Venus
The Trailblazing Path of Maria Mitchell

In the early nineteenth century nothing about the island of Nantucket made sense. It was simultaneously a hotbed of Quakerism and of the notoriously bawdy and violent whaling industry, a deeply conservative and God-fearing community that was at the same time at the forefront of gender equality in education and racial justice. For a few decades, before the collapse of the whaling industry and the ruin of Civil War, it was a place where just about anything could be expected to happen, and in 1847, it was here that American astronomy began, at the telescope of Maria Mitchell (1818–1889).

Mitchell was born in 1818 to a thoroughly unconventional Quaker family. Her mother had the reputation of having read every book on the island, and her father was one of America's first astronomers of note, as well as an amateur barrel-maker, bank president, schoolmaster and maritime instrument synchroniser. The parents and their multiple offspring all crammed together in a miniscule house with a makeshift observatory on the roof and, come nightfall, one child or other would gather on the perilously thin planking and aid William Mitchell's observations. One child, however, was his particular helper – his daughter, Maria.

Quakers had a long-standing tradition of equal education for girls and boys, so there was nothing so unusual about Maria's learning of the astronomical trade at her father's side. If anything, the community as a whole took pride in her evident scientific gifts, allowing her to become a member of their scientific society at the age of 12, and putting her in charge of the Athenaeum, the library for the island's intellectual set, at the age of 18. Science and poetry and history and languages were all taken very seriously by the Quaker community, and if somebody excelled, regardless of their gender, they were celebrated.

At the Athenaeum, Maria spent her time learning the mathematics which her father had never mastered and which would allow her to be a mathematical astronomer and not just a dedicated collector of data. Already in her teens, she was setting herself, and as a consequence American astronomy, down a road that would forever separate it from its Earnest Amateur past. While she would continue working with her father for the rest of his long life, her ability, to not only rigorously collect data but to interpret it to calculate elliptical and hyperbolic trajectories, would soon make her the senior partner in their endearing collective work.

She ran the library, helped synchronise the whaling industry's chronometers, and performed odd jobs for the navy, all very honourable and engaging work that would have seen her through to a quiet and modest retirement. But then, in 1847, on one of her nightly sky sweeps, she noticed something that had not been there before. It was a hyperbolic comet, and she was the first in the world to observe it. With natural Quaker modesty, she wanted to keep the discovery quiet, but her father, knowing what it might mean to the nascent American astronomical community, communicated Maria's data to Harvard, which rushed it to the British and Dutch astronomical authorities.

Maria's comet, together with her impeccable work in plotting its hyperbolic orbit, was the first real feather in the American astronomical cap, and made her an instant celebrity. During summer months, hundreds of tourists would descend on the Athenaeum for a chance to see her or collect her autograph. Newspapers ran stories about the Quaker astronomer who found a new heavenly body from her father's roof. More practically, a job offer came in paying a handsome $500 a year (her work at the library paid only $100, and her eventual professorship at Vassar only $800) to accurately plot the course of Venus for the purposes of navigation. That work would last nineteen years and make Maria Mitchell one of the few professional astronomers in the entire country.

It was, all things considered, a good time to be a woman in astronomy in America. The sciences were not professionalised as of yet, which meant low or no salaries and therefore low to no interest in claiming that intellectual ground as a solely male pursuit. Ideologically, maths and science in these pre-Darwin years were held to be relatively non-controversial paths to follow compared to the socialist and abolitionist rumblings of the humanities and so, while men were encouraged to study poetry and philosophy, women were positively thrust at the sciences in an amateur capacity. Collect flowers and stargaze by all means, as long as it keeps you from reading John Stuart Mill. For some time, Vassar's curriculum was scientifically the most robust in the country, and many women's colleges boasted more classes devoted to astronomy and biology than classics-leaning Harvard or Yale. Mitchell was the greatest and most esteemed fruit of these years of women's scientific education, and had the great misfortune to live to see the doors of science closed firmly against women after the Civil War.

During the war, the planning of Vassar (founded 1861 with the first students matriculating in 1865) as an exclusively women's college put a bold new idea at the heart of the educational system. Mitchell was invited to join the faculty as an astronomy professor at a salary of $1,500. She protested that she did not feel herself worth that much and they, somewhat shamefully, said, 'Great! How's about we pay you $800 instead?' She accepted, not aware that even the least experienced of male professors had been offered $2,000 a year. She was given a place to sleep in the school's observatory which turned out to be a cot set up in the office she was expected to teach students in, an indignity that lasted for ten years until they finally made up an old coal closet with a bed to serve as her room. Her widowed father lived with her in a small room provided for him in the Observatory's basement, Nantucket having more or less disintegrated under the social and economic pressures of the Civil War, and together they had a grand time acting as mentors to an aspiring crop of female astronomers.

In fact, Mitchell felt her role as educator so important that she took the radical step of diminishing her research projects so as to put more time into fostering the practical development of her students. They were welcome in the observatory at all hours, and only after the last had gone would Mitchell eke out some observations of her own. She still published regularly, but nothing of a particularly daring scope, her time given instead to devising ways to better teach her charges, to changing the classroom from a lecture-heavy format to one of practical applications and conversational interaction.

And Mitchell had one more concern, too. In 1873, Edward Clarke wrote a book, *Sex in Education*, which contained a troubling claim about the link between infertility and women's education. While becoming educated, Clarke argued, blood rushes from the uterus to the brain. The lack of uterine blood causes that organ to shrivel, creating first infertility and eventually effective hermaphroditism. Laughable today, this theory was taken very seriously at the time and, combined with the institutionalisation of science which made research jobs desirable for males, the press was on to push women out of scientific work. Mitchell fought back, serving as president of the Association for the Advancement of Women and working with leading feminist authors to organise a gathering of scientific data with which to refute Clarke's biological evidence.

While still famous and respected herself, Mitchell watched her pool of once talented, undefeatable students dwindle. Those who remained in astronomy faced familial persecution and a massively unsure future, the few available research jobs in their field being handed out to males while women were relegated to the role of photographic plate computers. Frustrated at the sight of an astronomical establishment which she helped put on the map turning so resolutely away from the eager spirits who had guided its first steps, and seeing nothing but further decline in the future, Mitchell retired from Vassar in 1888, and was dead a year later, carried off by a mysterious mental disease – possibly the same heart-breaking dementia that had taken her mother's life years before.

She was hailed in the papers for her life's work even as the male astronomical community efficiently closed ranks in the space left empty by her absence. Mitchell's legacy would be valiantly continued at Vassar by her student Mary Whitney, who published over 100 scientific papers, each a defiant cry that the spirit of Maria Mitchell was, contrary to all evidence and the best efforts of an entire academic discipline, still alive.

Further Reading

Maria Mitchell and the Sexing of Science by Renee Bergland is an at times brilliant, at times frustrating book. When it is at its best, detailing the surprising complexities of Nantucket Quaker life, and Maria Mitchell's place within both it, and the larger ante-bellum American scientific community, there are few books better. When it broadens its scope, however, it tends to lose focus and say regrettable things, from the thoroughly incorrect and insulting portrayal of Ellen Swallow Richards' foundational contributions to chemistry and ecology as non-rigorous and uninspired dawdling, to the less maddening but no less wrong portrayal of the bawdy Christina of Sweden as an 'ascetic'. Still, those moments are a small price to pay for the new vistas of insight into American science that Bergland offers.

Chapter 9

The Secrets Stars Keep
Lady Margaret Huggins, Pioneer of Spectral Photography

The history of science boasts a robust roster of Vanished Women, wives and sisters who did the work of full partners but received the recognition of mere Lab Helpmates (if they received any recognition at all). Each discipline has its share of vanished women, but in the early history of modern astronomy, as we have seen, they are absolutely legion. Great astronomers leaned on women as co-observers, analysts, technicians and theorists, but when it came time to write up their mutual results, it was the rare fellow indeed who placed anything but his own name under the title.

Two of the great exceptions to that trend elegantly exchanged the baton of life mid-nineteenth century: comet hunter extraordinaire Caroline Herschel (1750–1848), sister to astronomy legend William Herschel, and spectral astrophotographer Lady Margaret Lindsay Huggins (1848–1915), wife of stellar spectroscopy founder Sir William Huggins. Both women outlived their famous male counterparts, and spent much of their remaining energy loyally fighting to preserve their deceased's legacy instead of bringing attention to their own work, diligent acts of self-removal from the history of science that we are only now starting to undo.

Huggins, for her part, downplayed her importance so successfully that for a full century we knew nothing about her early life beyond a few stock anecdotes and hardly anything about the depth of her scientific contributions beyond the Victorian role of 'dutiful helpmate'. It took William eight years to acknowledge her co-equal role in their laboratory by placing her name on the reports of their joint work, and after his death she never attempted to outline her own role in the technical and theoretical development of their pioneering work in spectral astrophotography.

But she DID leave behind notebooks, and the story they tell is decidedly at odds with the Victorian face the Hugginses showed to the British astronomical community. We still know virtually nothing of her life before she married William. The two main facts that seem well established are that she picked up her devotion to astronomy from her grandfather and from popular science stories in Christian children's periodicals, and that she took an early interest in the emerging art of photography.

For a woman of the era, photography offered a socially approved way to engage in chemical experiments (mid-nineteenth-century photo development involved solutions and apparatus of almost alchemical grandeur), learn optics and the technical care of fine machinery, and develop her artistic abilities. Banned from the halls of experimental science, a significant number of young British women found their itch for research at least nominally scratched within the confines of a private darkroom. Young Margaret Lindsay Murray did not know it at the time, but her gift for nuanced photographic experimentation would be just the skill she needed to contribute fundamentally to her first intellectual love, astronomy.

As a child, Margaret had read about William Huggins' foundational work in stellar spectroscopy in one of her favourite magazines. Huggins was an amateur astronomer who had jumped early on the promise of spectroscopy (by which light from a star is spread out and examined for lines characteristic of different elements) to wrest from stars the secrets of their chemical composition. Pushing the science further, he had been the first to use Doppler Shift in a star's spectrum as a way of figuring out whether that star was moving towards or away from Earth, and how quickly, thereby giving humanity a moving, breathing picture of the motion of those seemingly still stars above.

His work had been so important that the Royal Society took the unprecedented step of spending £2,000 to set up an observatory in his home to allow him to continue and deepen his studies. A completely self-taught man, he has the distinction of being the Royal Society president possessing the absolute least amount of formal schooling. A brilliant amateur and outsider, he carried on a robust programme of investigation into the Sun's corona, the spectrum of planetary nebulas, and a host of other passing fancies besides.

To a young astronomy enthusiast, he was the stuff of heroes – an independent man blazing his own way along a path that compelled the stars themselves to sing their secrets. And so, when William and Margaret met at last it was perhaps a foregone conclusion that the famous stargazer and the star-struck astronomy and photography enthusiast would be drawn to one another.

They married in September of 1875.

She was 27.

He was 51.

They would spend the next thirty-five years of William's life together, united in their curiosity about what more spectroscopy might reveal about the structure of stars and other celestial phenomena. She came at precisely the right moment in William's career. Spectroscopy requires minute comparisons of the bright and dark lines of terrestrial and stellar spectra, comparisons that can at best only be roughly communicated in pen and paper representations jotted down mid-observation. Far better to harness the power of photography to create permanent records of each spectrum that can be analysed and compared at one's leisure, but what types of plates should one use, at what exposure times?

William, spectroscopic master though he was, had little knowledge of the science of the photographic techniques he would need if his laboratory was to stay competitive in an increasingly clustered scientific discipline. Margaret's expertise proved crucial in allowing him to turn the corner on the second phase of his career, and soon she was handling technical aspects of all the observatory's apparatuses and striking out on research projects of her own, while simultaneously lending crucial aid, insight and interpretation to his own work.

Together, they would pursue a varied and eclectic research programme that included attempts to photograph the solar corona without the benefit of an eclipse by using violet light filters, investigations of the spectrum of radium's strange and wondrous blue glow, the discovery of doubled helium lines in the spectrum of the Sun, and a prolonged fight against the almost bottomless academic aggression of Norman Lockyer as he attempted to foist his Basically Everything Is Meteorites theory of the galaxy on the astronomical community via his journal *Nature*. Some researches were crowned with success, and others crumbled under the weight of insufficient tools or theoretical structures, but what nobody could doubt was the solidity and reliability of the work coming out of the Hugginses' laboratory.

William and Margaret Huggins, by virtue of their careful cultivation of a younger generation of astronomical talent, and boundless curiosity untempered by old age (they began their radium studies when William was crossing into his eighth decade), stayed at the centre of astronomical research until the sheer scale of professional astronomy pushed them at last to retirement. Huge endowments for massive telescopes in America, and the growth of spectroscopy in university programmes, made the tight craftsmanship of the Hugginses appear increasingly quaint as the years went by. Eventually, they decided to return the telescope that William had been loaned by the Royal Society so many years ago, standing hand in hand as they watched the box bearing its object glass closed and taken away, leaving behind an empty room where once was the site of their greatest work together.

William died in 1910 at the age of 86, and Margaret devoted the rest of her life to correcting inaccuracies in the official accounts of his life and work, and ensuring that his equipment and notes found safe homes. Securing his legacy became the object of her remaining years, but then of course, when her time came in 1915 at the hands of cancer, there was nobody to do the same for her. She died the quintessential Victorian helper only to be reborn a century later as an innovator and technical master in her own right, steadfast in adversity, relentless in the search for truth, Lady Margaret Huggins.

Further Reading

A great debt is owed to Barbara J. Becker, whose *Unravelling Starlight: William and Margaret Huggins and the Rise of the New Astronomy* (2011) went back to the Hugginses' original notebooks and letters to present not only previously neglected aspects of William's research, but to bring about an entire re-evaluation of Margaret's role in his work. It is meticulously researched, engagingly written, and inspiring on multiple fronts.

Chapter 10

Making Spectroscopy Hip
Agnes Mary Clerke at the Nerve Centre of Nineteenth-Century Astrophysics

For a solid century and a half, from Mary Somerville's *The Mechanism of the Heavens* of 1830 to Helen Sawyer Hogg's final *With the Stars* column in 1981, it was largely from women that the English-speaking world learned its love of the night sky and its inhabitants. Somerville (1780–1872) had brought continental mathematics and celestial mechanics to an England that had been experiencing a heavy intellectual drought since the time of Newton, and Hogg (1905–1993) was for decades the steady and reliable popular source for enthusiasm about backyard astronomy and the cosmos beyond, but filling the void between them was perhaps the most important figure in the coordination, advancement and popularisation of nineteenth-century astronomy: Agnes Mary Clerke (1842–1907).

She lived during the tectonic shift of astronomy from a concern of enthusiastic, privately funded amateurs to the state-funded, big equipment, mountaintop observatory profession of the modern era, and was one of a handful of people not to get swallowed and obliterated in the transition. You do not have to look very hard to see why. She was a master of languages, and knew Latin, Ancient Greek, German, Italian and French on top of her native English. She once learned Portuguese in six weeks because she had some technical articles in that language she wanted to reference.

Apparently, that is something people can do!

That easy command of the world's major past and present languages of science made her the foremost historian of astronomy's past as well as the most gifted, widely read and internationally connected chronicler of its rapid late-century advancement.

She read everything and kept connected with everyone. Pickering at the Harvard Observatory, Gill at the Royal Cape Observatory, Hale at Yerkes, Wolf at Heidelberg, Holden and Barnard at Lick – these were *the* great names at *the* great observatories, and all of them checked in regularly with Clerke as their point of contact with the detailed goings on of the larger astronomical community. She received results from the world over before they were published and acted as the spokesperson for a field rapidly growing more complicated than the popular astronomy periodicals could simply convey. She saw clearly the value of spectroscopy (using the tell-tale elemental lines of a star's light to identify star composition and direction of motion) to revolutionise how astronomy was done, and promoted it with eloquence and infectious passion.

It was a calling honestly come by – her father was an enthusiastic amateur astronomer who used precise astronomical measurements to calibrate the main clock of the town where he worked as a banker. His shelves were filled with texts on astronomy and mathematics and he devoted himself to personally educating all of his children, giving them a grounding in science, mathematics, the classics and history that far outstripped the standard school curriculum for boys, not to speak of its titanic improvement over the Knitting And Dancing curriculum that British girls of the age could expect as the summit of their education.

After moving from Skibbereen to Dublin at the age of 19, Agnes added piano playing to her growing list of abilities, and would eventually meet and perform for Franz Liszt and study under a student of the legendary Berlin Philharmonic conductor Hans von Bulow. Making music was to be a lifelong love of hers, though she rarely overcame her shyness to do so for others. Meanwhile, her brother Aubrey studied astronomy at Dublin's Trinity College and brought home the newest developments in the field to be eagerly devoured by the star-mad Clerke clan.

In 1867, she had a chance to witness the birthplace of modern astronomy first hand when her family moved to Italy, there to remain for a decade. Agnes absorbed languages and history and took particular advantage of the open libraries to research the Italian masters of the cosmos, chief among them the complicated story of Galileo Galilei. The notes she took and sources she investigated stood her in good stead when she returned to England and began, with her sister, a career as an author. The calibre of her research and depth of her sources, combined with a native gift for writing, opened doors immediately to all the best periodicals of the day. In a story we do not often get to tell in Women in Science, instead of spending decades clawing for slivers of recognition, she enjoyed the sight of journals falling over themselves to feature her writing.

She made easy friends with editors, publishers, and members of the tight-knit international group of astronomers who were at that very moment working a revolution, and produced a stream of deeply researched articles about the modern state of science, as well as its deep past, totally unique in the science literature of the age. This was the time when astronomy was making the great spectroscopic shift, spreading out the spectra of starlight to determine what atoms were present and, after the application of Red and Blue Shift, what direction stars might be travelling relative to the Earth. With one method, astronomy went from the painstaking but relatively robotic categorisation of points in the sky to a deep knowledge about the chemical constitution and likely motion of those points, and Clerke was there to document these exciting early steps.

She wrote for *The Edinburgh Review, Observatory, Nature, Knowledge, The Dictionary of National Biography,* and the *Encyclopedia Britannica*, dense articles about new scientific developments culled from her network of contacts at the world's major observatories and her constant review of the scientific literature in four different languages. Keeping up with her extensive correspondence, visiting foreign observatories, entertaining visiting astronomers, and honouring her journal deadlines consumed masses of time, but somewhere in the midst of it all she also managed to write four books that, between them, both immaculately summed up the past astronomical age and gave shape and direction to the coming one.

A Popular History of Astronomy in the Nineteenth Century (1885) was her first book-length summation of astronomy's shift from star counting to star analysis, and it was, in spite of its unusual wealth of detail, a popular bestseller that went through four editions in two decades and long remained (and in some universities, remains) the standard history of the subject. She followed it up with *The System of the Stars* in 1890, which sought to sum up all the leading knowledge and theories about the nature and formation of stars and nebulae, and ultimately the universe itself. She was repeatedly honoured for the contribution these books represented to the cause of science – works of enough depth for professional astronomers to learn more about the breadth of their field, yet written clearly enough for amateur enthusiasts to follow and take inspiration from.

But it was her third book, *Problems in Astrophysics* (1903), that saw her breaking new ground. After years of observing, communicating and synthesising information from the world over, she felt it was time to put her storehouse of knowledge to positive use by setting out a definitive account of open questions to be pursued, along with suggestions about how the new wave of observational technology could be employed to answer those questions. She had been long privately pushing her astronomer friends to adopt various projects with regard to variable stars and peculiar spectra, but here at last she set a towering programme of research before the astronomical community. It was an accomplishment of such obvious importance that the Royal Astronomical Society had no choice but to break its over sixty-year ban on honorary memberships for women to bestow one on her.

All of her books were received with almost unilateral critical praise and popular appreciation, with Europe's and America's most esteemed astronomers reviewing each in words of unchecked admiration. She was seriously consulted for her opinion on astronomical problems by the age's most significant scientists, and her work would continue to be referenced in scholarly circles for a century after its original publication. Cripplingly shy of social encounters and public speaking, she never sought or accepted a research or academic position but kept to what she enjoyed best: sitting in her study and letting her brain reach out to the world through the printed word that had been her constant companion since her father placed book after book in her young hands and urged her to use them to explore the history of man and universe.

She died in 1907 of influenza, and the astronomical community mourned the loss as one in tribute papers and obituaries. She was the greatest synthesiser of England's great age of self-taught astronomy enthusiasts, and her death closed the book on a heroic era of daring expeditions and backyard observatories that, through sheer chutzpah and ingenuity, pushed us into a new understanding of what astronomy could do and say. She gave astronomy a sense of its fertile past and pushed it into a full appreciation of its coming potential, and took the rest of us along for the ride. And all it took was a pen, some books and some paper – and a mind ready at every moment to learn something new.

Further Reading

Mary Brück was the great historian of English and Irish women astronomers of the nineteenth century, and her book, *Agnes Mary Clerke & the Rise of Astrophysics* (2002), is the place to go for Clerke, but, if you can find it, a recommended follow-up is her *Women in Early British and Irish Astronomy: Stars and Satellites* (2009) though finding a reasonably priced one may be difficult.

Chapter 11

'The Somewhat Nerve-Wearing Experience'
How Sarah Frances Whiting Changed the Course of Women's Scientific Education

In the 1860s and 1870s, women's access to advanced astronomical instruction in the United States took a sudden and majestic turn for the better, even as women's prospects to enjoy a career as astronomical researchers were taking a turn for the worse. In 1865, Vassar's first students matriculated, and found the legendary Maria Mitchell at the helm. Shortly thereafter, in 1869, Swarthmore opened its doors, with Susan Jane Cunningham, Mitchell's former student, taking charge of astronomy instruction there for the next four decades. In 1875, two important women's colleges opened, Smith College (the observatory of which would soon come under the direction of Mary Emma Byrd) and Wellesley College, which featured the powerful presence of Sarah Frances Whiting (1847–1927) as the head of both its revolutionary physics and astronomy programmes.

One of the great problems that immediately arose for the founders of these new colleges was how to staff them with trained and professional women. Part of the reason that the colleges were formed was to meet the American dearth in higher education for women. Not many women had the chance to pursue a college degree, and of those that did have the opportunity, many did not make it to the end of their studies (in Whiting's Wellesley, for example, of a first class of 246 students, only eighteen ultimately received their degrees) due to family or financing problems, and of those that *did* get their degrees, many married soon thereafter which, in nineteenth-century America, meant the end of their hireability as teachers, with many states having laws on the books declaring it illegal for a married woman to hold a teaching position. Staffing four ambitious colleges with the remaining talent was a challenging matter, but fortunately for the Fowle Durants, the patrons of Wellesley, the perfect individual for their new physics department had just recently emerged from the women's academic pipeline, such as it was in that era.

Sarah Whiting was a graduate of Ingham University, which upon its founding in 1835 became the first women's college in New York, and the first women's fully chartered university in the United States. By 1892, Ingham had closed, but fortunately Whiting attended during its more robust middle years, earning her degree in 1865. For the next decade she taught at a Brooklyn secondary school, as graduate study opportunities for a woman in that era were distinctly thin on the ground. Well might she have expected to spend her entire career in that position, or in a series of similar ones, all leading to a semi-comfortable if thread-worn retirement. She had not, however, given up entirely on the prospect of expanding her horizons, and attended whatever lectures she could in what would become a lifelong habit of seeking out the Latest Thing in scientific developments and learning all she could about it directly from those working on the cutting edge of research.

In 1876, she was chosen to become Wellesley's first professor of physics, and in preparation for that work, she attended lectures at the Massachusetts Institute of Technology by none other than Edward Charles Pickering, whose name will recur many times in this volume as a result of his assumption of the directorship of the Harvard Observatory in 1877. Pickering had a novel idea about how physics should be taught, as a hands-on experimental course where students gained experience as undergraduates handling scientific equipment and carrying out their own research projects. Whiting resolved that Wellesley's physics programme would continue and deepen this example, as she took charge of stocking the nation's second undergraduate laboratory (and the first accessible to women), and designing engaging curricula around it.

Whiting's methods proved so successful that she was soon given broader responsibilities, and started teaching astronomy courses in addition to her physics classes, beginning in 1880. Teaching astronomy, she employed similar methods to those she had used in her physics courses, and her students found themselves, as undergraduates, learning real astronomical techniques, particularly in the booming field of stellar and solar spectral analysis, in which pursuit dozens of women would soon be employed (though at near-starvation wages) in observatories across the country. Until 1900, the only telescope Whiting had available to offer students was a 4-incher they used to plop on the roof of College Hall, but in 1896 a Mrs J.C. Whitin wrote to Whiting about a good deal to be had on a 12-inch refracting telescope, and soon proposed to purchase not only the telescope for Wellesley, but a building to house it as well, to be built of white marble. In 1900, the Whitin Observatory opened, with not only the 12-inch telescope, but a micrometer, polarising photometer and a variety of spectroscopes, and it was only natural that Whiting was named its first director.

In addition to her work as a professor, observatory director, laboratory manager and department head, Whiting was also an author, popularising the new experimental approach to astronomy pedagogy through a number of articles, as well as the textbook *Daytime and Evening Exercises in Astronomy, for Schools and Colleges* (1912). She retired from teaching in 1912, and from her position at the Whitin Observatory in 1916, after some forty years of service to Wellesley, in which time she saw a number of her students rise to important positions in the sciences, including Annie Jump Cannon, whom we shall soon meet, and who put the spectrographic training she learned with Whiting to truly epic use during her time on the Henry Draper Catalogue project at Harvard; the physicists Isabelle Stone and Louise McDowell; and the meteorologist Grace Evangeline Davis.

Speaking of her experience attending numerous conferences on new developments in physics and visiting laboratories all over the world, Whiting once mentioned 'the somewhat nerve-wearing experience of constantly being in places where a woman was not expected to be, and doing what women did not conventionally do'. Undoubtedly, the experience for Whiting was something more than nerve-wearing, as many of her male colleagues made little secret of their contempt for women attempting to do research in science, but thanks to her, and her innovative methods for putting practical experimental experience into the hands of her undergraduates, those that followed her could enter those male spaces, where they were not expected or often welcomed, stroll to the nearest spectrometer, and show just what stuff they were made of.

Chapter 12

'Bordering on the Marvellous'
The Astronomical Menagerie of Williamina Paton Fleming

When William Huggins became the first human to examine the spectrum of a star in 1864, he could little have imagined the magnitude of the project he had brought to life. With the analysis of stellar spectra, astronomy took a bold leap forward from comet hunting and position plotting to a thoroughgoing chemical analysis of the stars which had the potential of unveiling wholly new models of the cosmos. In the United States, the Harvard Observatory under Edward Pickering (director 1877–1919) undertook a massive project in 1886 to photograph and categorise the spectral contents of the night sky, an investigation that would generate first hundreds, then hundreds of thousands, of photographic plates. Generating those plates was a simple enough matter, but the real question was, who was going to *analyse* them?

Put another way, after the official astronomers finished amassing their photographical observations at night, who was going to put in the hours of work during the subsequent day, poring over those plates, recording the significant features of their spectra, comparing them to previous measurements, and categorising the underlying objects based on variability, composition, colour and age? Pickering could have hired more men for the task, but men were expensive. The answer, it seemed, was women, who could be hired for 25 cents a day, and some of whom would even work for free just to gain expertise in working with real, current astronomical data.

In addition to the issue of who was going to perform all the analytical work, was the question of who was going to organise and oversee it all. Who could be trusted to take this mind-boggling amount of raw data – for the time – and create administrative structures and procedures to make sure that standardised and reliable categorisations would be attained, regardless of which woman happened to be analysing a particular plate? As luck would have it, Pickering had the perfect person for that job directly at hand. Williamina Paton Fleming (1857–1911) had served as Pickering's housekeeper after a spell of bad personal luck, but her intelligence and organisation had been noted by Pickering's wife, who recommended that she be employed as an office assistant rather than as a household drudge.

It was a fateful decision – Fleming was an administrative and observational virtuoso with a profound memory, capable of discerning fine gradations of variation which she collected in her head, and organising the observed phenomena into the groupings that would become the categorisations she trained her team to recognise and record. She was, in many ways, the Linnaeus of astronomy, employing her experience and sharp recall to tame the spectral chaos being generated by the Harvard Observatory, and her skills at differentiation to note and document entirely new astronomical phenomena.

For the first twenty-two years of her life, however, there was little to suggest the powerful scientific figure she would become. She was born Williamina Paton Stevens in 1857 in Dundee, Scotland, to a family that descended from the Claverhouse Grahams, who traced their lineage back to the fourteenth-century king, Robert III, and included among their ancestors the first Viscount Dundee, John Graham of Claverhouse, who was a hero of the seventeenth-century Jacobite Uprising. Though rich in ancestors, Williamina's childhood was hardly one of lordly ease. Her father was a craftsman who died when she was 7 years old, leaving her, her mother and eight other siblings to get by as best they could. She aided the family purse with her brain, becoming a student teacher at the age of 14, and at 20 she married James Orr Fleming, a bank employee some sixteen years her senior.

The next steps in Williamina's story are hazy, but what is sure is that at some point, either before or after Williamina emigrated to Boston, Massachusetts, James abandoned her and her as-yet-unborn son, to look out for themselves. Some of Fleming's siblings lived in Boston, so she was not entirely without support, but she would need some regular employment with a new child on the way, and in 1879 she found the position as maid to Edward Pickering which would place her life on an entirely new trajectory.

Later that year, Fleming was moved to office work at the observatory as an occupation both more in line with her talents, and better suited in nineteenth-century estimations to a woman in the more advanced stages of pregnancy. The Pickerings even went as far as to help pay for Fleming to return to Scotland for her laying-in and childbirth. In recognition of all the Pickerings had done for her during her time of greatest need, Fleming named her child Edward Charles Pickering Fleming.

Fleming returned to the United States and her work at the observatory, leaving young Edward behind in the care of her mother. Upon her return in 1881, Pickering formally promoted Fleming from office worker to a full-time member of the observatory staff, with particular responsibilities for recording and systematising relative star brightness. In the course of this work, she noted over 300 'variable stars' which changed their brightness over time, and one type of which, the Cepheid variable star, would prove central to Henrietta Leavitt-Swan's revolutionary work of 1908 (to put this achievement in perspective, when Fleming began her work, fewer than 200 variable stars were known to astronomers). She later gathered 222 of these and provided reference data for their future tracking, an effort that was summarised by the author of her Royal Astronomical Society obituary as follows: 'Many astronomers are deservedly proud to have discovered one variable, and content to leave the arrangements for its observation to others: the discovery of 222, and the care for their future on this scale, is an achievement bordering on the marvellous.'

As marvellous as her achievements were, they were about to increase another order of magnitude still, as in 1886 the widow of spectral astronomer Henry Draper approached Edward Pickering with an idea to fund the spectral study of the night sky, a project known today as the Henry Draper Catalogue. The work of coming up with a more refined categorisation than the four Secchi classes that had been used theretofore, and then applying that system to the observatory's spectral data, largely fell to Fleming, who created a new system with twelve main and four auxiliary categories, labelled A–Q. The first edition of the Catalogue, published in 1890, contained 10,351 stars categorised primarily by Fleming from 28,266 spectra spread across 633 photographic plates.

It was a dizzying achievement for a pre-digital age astronomer, and it was not nearly at an end. In 1897, Fleming and Antonia Maury would publish an updated and expanded edition of the Catalogue, and in 1890 a second observatory was opened by Harvard in Peru to photograph the Southern Hemisphere sky, the data of which was analyzed by Annie Jump Cannon using a reduced and rearranged form of Fleming's categorisation system, the OBAFGKM system still employed today. The final volume of the Henry Draper Catalogue was published in 1924, by which time over 225,000 stellar spectra had been recorded and categorised by Fleming's team and its descendants.

But Fleming's achievements are not only about quantity and brute force perseverance but about sharp scientific intuition as well, whereby experience and insight combined to notice unusual phenomena, and focused research thereupon. Many one sentence biographies of Fleming focus on her as the discoverer of the Horsehead Nebula, which she had noticed on a photographic plate in 1888, and the credit for which was initially given to Pickering, a mistake corrected in the 1908 edition of the Draper Catalogue.

The Horsehead Nebula was one of fifty-nine nebulae discovered by Fleming, which is thoroughly astounding, but what I want to talk about even more is her discovery of white dwarf stars in 1910. Henry Norris Russell had asked Pickering to determine the category of a collection of faint stars he had been observing. Pickering agreed, and Fleming categorised most of the stars as class M, which consist of yellowish-red stars that make up 76 per cent of all stars in the universe and have the lowest temperature of any stars in the OBAFGKM system. (Our own Sun is a class G star; in case you were wondering.) One of the faint stars, however, 40 Eridani B, Fleming categorised as class A. Such stars have 3–4 times the temperature of M stars, and tend to burn white.

The result was a bombshell in Russell's ordered conception of the cosmos. Luminosity, which affects how bright an object appears to our instruments, is determined primarily by two factors – the object's radius, and its temperature. As a hot class A object, 40 Eridani B should have been relatively bright – its lack of luminosity, then, could only be explained by it having some manner of freakishly small radius, a new type of inhabitant in the cosmos, eventually dubbed a 'white dwarf' in 1922. White dwarfs occur when stars collapse that do not have enough mass to form a black hole. The result is an ultra-dense object the size of the Earth, but with the mass of a sun, in which fusion no longer happens, but which can still have high surface temperatures when it first forms. In later accounts of the discovery, Russell recognised Fleming as 'the first person who knew of the existence of white dwarfs'.

Having discovered over 300 variable stars, 59 nebulae, 10 novae (including the first ever nova discovered through spectral techniques, and which spawned powerful new tools for revealing future novae), the Horsehead Nebula, and the first white dwarf star, Fleming was a vastly respected individual in the astronomical community, which honoured her as much as its baked-in institutional inertia would allow. Pickering promoted her in 1899 to the position of 'Curator of Astronomical Photographs', a job created particularly for her to officially recognise the magnitude of her contributions, and the first titled position ever offered a woman at Harvard. In 1906, the Royal Astronomical Society made her an honorary member, the first Scottish woman to attain that distinction (the first women made honorary members were Caroline Herschel and Mary Somerville, both in 1835), and she was also a charter member of the Astronomical and Astrophysical Society of America, and a member of the Société Astronomique de France.

She worked steadily at her job, being paid 60 per cent of the salary a man in her position would have earned, supporting herself and her spendthrift son Edward while contributing foundationally to one of the early twentieth century's most monumental astronomical undertakings and pushing through the surgical procedures that pervaded her last two years, until she was hospitalised for pneumonia in the spring of 1911, ultimately passing away on 21 May 1911. The system she set into motion, however, remained in place after her death, resulting not only in the towering achievement of the Henry Draper Catalogue but in providing the space that launched the careers of some of the century's most estimable astronomical figures, including Annie Jump Cannon, Henrietta Swan Leavitt, Antonia Maury and Cecilia Payne-Gaposchkin.

Further Reading

Like many of the figures in this volume, Williamina Fleming has yet to receive a stand-alone volume devoted to her life and works, but she does appear often in works about the early Harvard Observatory, its computers, and key figures. Dava Sobel's *The Glass Universe* (2016) is devoted entirely to Pickering and the history of the women computers at the Harvard Observatory, and is a great place to start. You can also find her Royal Astronomical Society obituary in the 4th issue of the 72nd edition of the *Monthly Notices* (Feb 1912). And, if you want to do something really cool, you can check out the online Henry Draper Catalogue which has links to the thousands of stars recorded by Fleming, Maury and Cannon, with pictures and locations for each!

Chapter 13

Brief Portraits
The Great Nineteenth-Century Explosion

Elizabeth Juliana Sabine (née Leeves) (1807–1879) and Elise Otté (1818–1901)

One of the myriad ways women in the nineteenth century aided the cause of science was by employing their linguistic skills, combined with their scientific education, to produce translations of important international scientific works. Some books proved so popular to the general public that a number of different women produced their own individual translations, with one notable instance being the competing translations of Alexander von Humboldt's five-volume *Kosmos*, published between 1845 and 1862, which attempted to construct a harmonious account of the unity of the natural world that wove together knowledge from numerous scientific fields. Elizabeth Sabine, who as the assistant and tireless computer for her husband, the astronomer Edward Sabine, would deserve mention in this volume even if she had never translated a single word, published her translation of the first four volumes in 1849, though the title page, instead of bearing her name, carries the phrase 'Translated under the Superintendence of Lieut-Col. Edward Sabine, R.A., For. Sec. R.S.' What was important was not recognising the woman who actually did all the translating, but rather shoehorning in the name of the man who was 'supervising' her, making sure to get in as many of his distinctions and titles as possible. In fact, over the course of the entire four volumes, her name is mentioned only once – in the editor's preface.

Sabine's translation, while masterful, also suffered from a major defect – either Elizabeth or Edward deemed some of Humboldt's passages too controversial for the English religious taste, and made the decision to simply excise them from the text. The omissions drew criticism from the scientific community, even as they recognised the invaluable nature of Sabine's translation of the non-excised bits, leaving room in the national consciousness for the Anglo-Dutch Elise Otté to enter the arena with her unedited 1851 translation, which had the secondary virtue of being much cheaper than Sabine's. Though Otté did not, like Sabine, have a secondary career as an astronomical assistant, she did publish other translations of scientific works, and was at the centre of her stepfather Benjamin Thorpe's work to popularise classic Norse literature in England through translation.

Caterina Scarpellini (1808–1873)

In the late eighteenth century, some elements of the Roman Catholic Church, watching the general thrust of European civilisation passing them by, began pushing for a new approach to Catholicism which has loosely been termed (though, as with many modern academic coinages, there is little to no agreement about what it means, who it applies to, or the time period it covers) the Catholic Enlightenment. Part of that movement was a drive to make up for lost time in the sciences, and to present Christianity as complementary to the new scientific spirit of the age, rather than in the adversarial role it had played during the heliocentric battles of the sixteenth and seventeenth centuries. To this end, in 1778 the first astronomical observatory was established in Rome (the Observatory Caetani, financed by Francesco Caetani, Duke of Sermoneta), followed by the founding of the Caetani Academy, which in 1801 changed its name to the Accademia dei Lincei, which name was a callback to the spirit of Galileo Galilei's Accademia dei Lincei, which existed from 1603 to 1645. In 1816, the Pope himself got in on the science game, by establishing a chair in Sacred Physics (a title which has never failed to bring me great joy whenever I read it) which was first occupied by Abbot Feliciano Scarpellini, who was also a founding member of the new Accademia dei Lincei, and by founding the Campidoglio Observatory in 1829, of which Scarpellini was made the director.

Feliciano Scarpellini, then, was at the very centre of the Papacy's attempt to re-invent itself scientifically, and particularly in the effort to improve its centuries-long reputation as the bête noire of astronomical truth. It was a large job, and to pull it off he would need help, which would come in the form of his supremely talented niece, Caterina. In 1826, at the age of 18, she arrived in Rome to assist Feliciano, and over the next five decades she built a reputation as one of the continent's most accomplished scientific investigators. She made astronomical observations six times a day, and published summaries of those observations in Italian, Belgian and French journals, while also editing, from 1847, the *Corrispondenza Scientifica* which served as one of Rome's important sources for new scientific information. She was also engaged in meteorological observation, to which end she and her husband Erasmo Fabri established a meteorological station in Rome in 1856.

Her writings and theories covered a wide range of phenomena, from her 1864 chemical analysis of sand that had blown into Rome which placed its origins in the Sahara desert, to her treatises on Saturn's rings, to her compilation of the first Italian catalogue of meteors, to her hypotheses about the lunar influence on earthquakes, to her hydrothermal observations of the Tiber River (the first ever attempted), to her 1854 discovery of a new comet, to her insights into the smallpox epidemic that struck Rome in 1871, to her studies of geological and magnetic phenomena. She was, simply put, a scientific force, and was recognised as such by the Italian academic establishment, which made her a member of the Accademia dei Georgofili (founded in Florence in 1753), and by the government, which recognised her achievements in 1872, and erected a statue in her honour in the Campo Verano cemetery in Rome.

Mary Parsons, Countess of Rosse (née Field) (1813–1885)

Mary Rosse got things done. A devotee of discipline and labour, she reportedly ran her household with an almost military-like efficiency, evincing a gift for organisation that she would publicly apply in the service of astronomy as a driving force behind her husband, Lord Rosse's, great astronomical projects. She was the daughter of a wealthy estate owner, and married Lord Rosse, whose interest in astronomy matched her own, in 1836. She placed £12,000 of her private wealth (equivalent to £850,000 in today's currency, or just over $1 million at the current exchange rate) at her husband's disposal for constructing the Leviathan of Parsonstown, a massive 72-inch reflecting telescope constructed from 1839 to 1845, which remained the world's largest telescope until 1917. Mary documented the process of constructing the Leviathan in a series of historically priceless photographs. In addition to her role as primary financier, it is speculated by some historians that the countess also played a role in the development and planning of the project. She was, further, self-trained as a blacksmith, and was among the first generation of photographers to use wax paper negatives (which is a method of creating a transparent negative from a paper negative through the application of melted wax).

Historians of engineering in the audience might have been impatiently squirming in their seats throughout this whole paragraph, waiting to shoot up their hands and ask, 'Is Mary Parsons related to Katharine and Rachel Parsons, the founders of the Women's Engineering Society?!' And the answer, is YES! Their father, Charles Parsons, was one of the four out of Mary's eleven children who survived to adulthood, making her their grandmother.

Mary Ward (née King) (1827–1869)

One of the tragedies about Mary Ward is that the manner of her death is the thing people remember about her, rather than the content of her life. Yes, she was probably the first person ever killed by a motorised vehicle – great fact to have in your pocket for the next time you are on *QI* – but she was also an astronomer, naturalist, microscopist and artist who deserved more than the forty-two years upon this Earth she got. She was born in the Irish town of Ferbane to a scientific family that included her cousin William Parsons (Lord Rosse), whom we have already met as the builder of the 'Leviathan of Parsonstown', a 6-foot diameter reflecting telescope that reigned from 1845 to 1917 as the largest telescope in the world and that boasted a 5-inch thick, 3-ton mirror cast personally by Parsons. In fact, Ward was the first person to look through the Leviathan when it became operational.

One day, the astronomer James South saw her sketching insects, using a magnifying glass to attempt to view the smallest details in order to capture them in her drawings, whereupon he remonstrated with her father to purchase her a proper microscope. This her father did, and so kindled Ward's lifelong interest in microscopy, a topic she wrote extensively about in the books *A Windfall for the Microscope* (1856), *A World of Wonders, Revealed by the Microscope* (1857, reprinted eight times over the next two decades), and *Microscope Teachings* (1864). She did not forget her familial roots, however, amidst her microscopic musings, and maintained her interest in astronomy. Along with Queen Victoria and Mary Somerville (who we will meet in detail in the Mathematics volume of this series), she was one of only three women on the Royal Astronomical Society's mailing list, and in 1859 wrote the book *Telescope Teachings*.

In an interesting bit of matrilineal continuity, 110 years after Mary Ward's death, her great-granddaughter Lalla Ward carried on the family tradition of bringing the wonders of space to the people when, in 1979, she assumed the role of Romana on the television programme *Doctor Who*.

Elizabeth Brown (1830–1899)

Elizabeth Brown was the daughter of a wine merchant whose passion was amateur astronomy. Whenever he could, Thomas Brown made meteorological observations, and recorded sunspot activity, and brought his daughter Elizabeth into these daily rituals, teaching her how to use the telescope and record her observations. Thomas's poor health, however, kept Elizabeth close to home, and far from a formal education, so that it was not until his death in 1883 that Elizabeth, age 53, was able to begin her own path.

Her years of observation with her father made astronomy an obvious choice for where to direct her energies, and she soon joined the Liverpool Astronomical Society, which had been founded in 1881 and was already developing an excellent reputation for amateur astronomical pursuits. Later, she was one of four women on the founding council for the British Astronomical Association, which became a critically important refuge for women astronomers in the long years waiting for the Royal Astronomical Association to admit women. (The other three women on that council were Margaret Huggins, Agnes Clerke and Agnes Giberne.) From 1890 to 1899 she was director of the BAA's solar section, a perfect fit for the woman who had spent the better part of five decades recording solar phenomena. In 1892, she was, along with Annie Maunder and Alice Everett, proposed for fellowship in the Royal Astronomical Association but, like Isis Pogson (see below) six years earlier, was rejected.

Brown was a prolific writer, author of two astronomical memoirs: *In Pursuit of a Shadow* (1887) and *Caught in the Tropics* (1890) (of which the former is still available in facsimile editions), as well as dozens of articles on solar astronomy written for the *Journal of the British Astronomical Association*, *Nature*, and *The Observatory* through the 1880s and 1890s. In 2020, a play about Elizabeth Brown, *The Making of Miss Brown*, by David Elder, debuted at the Stroud Theatre Festival.

Thereza Dillwyn Llewelyn (1834–1926)

In the heady early days of photography, while many employed the new technology to indulge the inexhaustible interest of humans in looking at themselves for prolonged periods, a few saw it as a powerful new tool in the service of science. Thereza Llewelyn, born in the Welsh county of Glamorgan, came from a family of photography pioneers, including her botanist father John Dillwyn Llewelyn, and her aunt, Mary Dillwyn, who was one of the first women in Wales to take up an interest in photography. Llewelyn took to astronomy from an early age, and to encourage it her father built her an observatory for her sixteenth birthday. Together, they worked on how best to incorporate photography into astronomical observation. This was a tricky proposition – in the early days of photography it required upwards of half an hour of exposure time to create an image, plenty of time for the Earth's natural rotation to move a celestial object out of frame entirely. Llewelyn and her father compensated for this in the days before electronic synchronisation machines by physically counter-rotating the telescope ever ... so ... gently to keep their subject (generally the Moon, of which they made some of the earliest photographs in the 1850s) in frame without smearing its features.

Mary Ann Albertson (née Mitchell) (1838–1914)

A cousin of Maria Mitchell, Mary Ann Albertson was, from 1904 to her death in 1914, employed by the Maria Mitchell Association (formed in 1902 in Nantucket to honour the memory of Maria Mitchell and carry on her educational and research goals) as curator of its observatory and department of botany, and as librarian to the Maria Mitchell Science Library. She created a special herbarium of Nantucket flowers in honour of her cousin, who possessed a love of flowers rivalling her devotion to the stars.

Elizabeth Isis Pogson Kent (1852–1945)

Isis Pogson (as she is most commonly known) served as astronomical assistant to her father, English astronomer Norman Pogson (1829–1891), at the Madras Observatory of India from 1860 to his death in 1891, and then continued on at that institution until its closing in 1898. She also worked as a computer for the observatory from 1873 to 1898, earning an annual salary of 150 rupees, which was roughly that of a cook. In 1881, she added to these duties that of meteorological superintendent and reporter to the Madras government. Her father proposed her for membership in the Royal Astronomical Society in 1886. It was the first time a woman was nominated as an actual fellow rather than an honorary member (Anne Maunder would not be proposed until 1892), and so the case was referred to legal experts, who stated that, as the 1831 society charter only ever employed the pronoun *he*, therefore it must be the case that women were forbidden from being fellows of the RAS. That iron-clad reasoning remained the rule until 1916, when the first women were admitted to the RAS. Pogson then had to wait four further years, until 1920, for her own admission, some thirty-four years after she was first proposed.

Susan Jane Cunningham (1842–1921)

What Maria Mitchell was to Vassar, Susan Jane Cunningham was to Swarthmore – the guiding force of its astronomical programme, which she saw from its foundation in 1869 through to her retirement four decades later, in 1906. Born in Maryland to a Quaker family, she attended Vassar in its second year of operation (1866) to take classes from Maria Mitchell, and regularly for the next thirty years took every opportunity she could to visit other institutions during her summers and learn from what they were doing there, including the Harvard Observatory (1874, 1876), the Princeton Observatory (1881), the Cambridge Observatory (1887), and the Greenwich Observatory (1891). She was a lifelong teacher, and her great opportunity at using her skills as an educator to lift up a generation of curious minds came in 1869 when Swarthmore (then a Quaker co-educational institution) opened its doors, and she was selected to form the mathematics and astronomy departments. She was made a professor of mathematics in 1871, and was made chair of that department in 1888, the same year she was given the go-ahead to plan Swarthmore's very own observatory, which came to be known as, simply, the Cunningham Observatory. She was also given, again in 1888, an honorary doctorate by Swarthmore.

Though most of her time was occupied with running two departments and seeing to the needs of her students, Cunningham also somehow had the capacity to break new ground in a number of institutions, becoming a founding member of the British Astronomical Association organised by William Maunder in 1890 (of which we will have much more opportunity to discuss below), a member of the first group of women admitted to the New York Mathematical Society (today the American Mathematical Society) in 1891, and a fellow of the American Association for the Advancement of Science in 1901.

Still, it is for her time at Swarthmore that she will probably always be remembered. In the words of that institution's president upon her retirement in 1906, 'She has loved Swarthmore more than her own life, of which she has unsparingly given. She has in season and out of season been ready not only to serve the College but to help individual students by giving them her advice, her time and in numerous cases her money.' When she died in 1921, her funeral was attended by Pennsylvania governor William J. Sproul, her former student.

Mary Watson Whitney (1847–1921)

When Maria Mitchell retired in 1888, one year before her death, she left behind some massive shoes to fill. Whoever succeeded to her position at the Vassar Observatory would inherit not only a position that had been occupied by America's first professional astronomer, and first ever woman professor of astronomy, for the last quarter of a century, but also the status as role model to an entire nation's prospective women astronomy students who, having lost their particular hero, were casting about for a new figure to inspire and motivate them. Fortunately, the ideal person was at hand, one of Maria Mitchell's very first students, and a figure tailor-made to carry on wearing the crown of her former teacher – Mary Whitney.

Whitney possessed a gift for mathematics that was noticed and encouraged from her youth, and was born at very nearly the perfect moment to take advantage of developments in higher education for women that were all but unknown to an earlier generation. Graduating from high school in 1864, she had to wait but one year for the opening of Vassar in 1865, where she formed one of the inaugural class of 353 students. She was one of the standout students from the beginning, referred to by fellow students as 'Pallas Athene, Our Goddess of Wisdom'. She also formed one vertex of 'The Hexagon', a group of six astronomy students who took a particular interest in the events organised by Maria Mitchell. Graduating in three years, Whitney returned home to support a family weakened by the loss of her father in 1867 and brother in 1868.

You cannot keep a good mathematical astronomer down, however, and by 1869 she was attending lectures by the Harvard mathematician Benjamin Pierce on quaternions, which are like complex numbers, but contain two more terms, a 'j' and 'k' term, beyond the usual 'i' term. Created in 1843 by William Hamilton, quaternions were still relatively exotic mathematical objects in 1869, with intriguing applications for three-dimensional rotations. Whitney enjoyed the lectures, and was invited by Pierce to attend a private class he was giving on celestial mechanics. Shortly thereafter, in 1872, Whitney received a master's degree from Vassar, which she followed up with studies at the University of Zurich from 1873 to 1876.

For all of her education, however, opportunities to apply it were few and far between, and for five years from 1876 to 1881 Whitney entered a dark and unsatisfying period, back home teaching at Waltham High School, where only a fraction of her mathematical or astronomical know-how was employed at any given time. Fortunately, in 1881 she received the call from Maria Mitchell to be her assistant, a role that became all the more important as Mitchell's health began to fail. By 1887, when Harvard Observatory offered Whitney a research position, she felt she had to turn it down in order to be of service to her old teacher in her difficult final years.

Mitchell retired in 1888, and Whitney was the clear choice for her successor. While she dearly loved Mitchell, Whitney did recognise areas where astronomical studies at Vassar could be improved, particularly as regards inaugurating a rich research programme. Under Whitney's watch, 102 articles were published by Vassar astronomy students and faculty, providing the next generation of astronomers with skills that would prove valuable in their future careers.

Whitney served for twenty-two years as the head of the Vassar Observatory, until in 1910 a cerebral haemorrhage paralysed the left half of her body. She took a leave of absence from Vassar, hoping for improvement, and when that did not come, she retired in 1915, succeeded at the observatory by her former student Caroline Furness (see below).

Mary Emma Byrd (1849–1934)

Another prominent figure in mid-nineteenth-century astronomical education, Mary Byrd was the child of noted abolitionist Reverend John Huntington Byrd, a strict Puritan who managed a station in Kansas for the Underground Railroad. She attended Oberlin College in 1871 but left before graduating, ultimately receiving her bachelor's degree from the University of Michigan in 1878. She was a principal at Indiana High School from 1879 to 1882 until she found a position at, you might have guessed it, the Harvard Observatory under Edward Pickering. Unlike Florence Cushman or Anna Winlock, however, who spent decades under the protective, if not always robust, umbrella of Pickering's Harvard, Byrd only remained for two years. For the next two decades, she mixed teaching with observatory work with her graduate studies, acting as first assistant at the observatory of Carleton College from 1883 to 1887, as observatory director at Smith College from 1887 to 1906, and as professor of astronomy there from 1898 to 1906.

At Smith, she wrote the popular practical textbook *Laboratory Manual in Astronomy* (1889) and carried on research using stars of known position to categorise the position of comets. All was going well until 1906, when she discovered that Smith College received funding from the Carnegie and Rockefeller foundations. To her Puritan upbringing, the idea of a college sullying itself with money derived from industry was insupportable, and she resigned in protest. Thereupon she returned to Kansas, and settled into a semi-retirement, writing popular articles on scientific topics and a second textbook, *First Observations in Astronomy: A Handbook for Schools and Colleges* (1913), thereby cementing her reputation as one of the most important forces in the development of college-level American astronomy instruction.

Henrietta Janssen (fl. 1880s)

Of all the figures in the last 200 years portrayed in this volume, Henrietta Janssen is perhaps the one most shrouded in mystery. We do not know her original last name, and for the longest time we did not even know her first, with sources from as recently as thirty years ago having to rest with 'Madame Janssen' as her identity. Her birth date, death date, education, all rest as tremendous blanks next to the one salient fact we do know about her, which is that she was married to Pierre Jules Janssen, a French astronomer who is among those with a strong claim to having first discovered helium. In this role, she is recorded as having been instrumental in the planning of annual expeditions to the Mont Blanc observatory constructed by her husband in 1893, and of playing a similar role in his international astronomical expeditions. There are apparently 200 of her letters existent at present in the archives of the Institut de France, so if anybody out there gets some time with them, let me know and maybe we can clear out some of these cobwebs in the life of Henrietta Janssen.

Lidiya Petrovna Tseraskaya (1855–1931)

The Astrakhan-born Tseraskaya was educated at St Petersburg's Teacher's Institute, and spent her career at the Moscow Observatory, where she was responsible for the discovery of some 219 variable stars. In 1884, she married the astronomer Vitold Tserasky, who in 1890 became director of the Moscow Observatory.

Anna Winlock (1857–1904) and Selina Bond (1831–1920)

Williamina Fleming was not the first woman employed at the Harvard Observatory to perform astronomical computations and inspect photographic plates. When she arrived in 1879, there was already working at Harvard a pair of women with deep connections to the Observatory. Selina Bond was the daughter of the Observatory's first director, William Cranch Bond (director 1839–1859), and sister to its second director, George Phillips Bond (director 1859–1865). After the death of her father and brother, Selina Bond found herself in almost perpetual financial difficulty, and so when Pickering assumed directorship of the Observatory in 1877, one of the first requests he received was from Bond for computer work, work she continued at home when she was no longer able to physically come to the Observatory on account of her poor health. One of the projects she devoted particular time and energy to was in aiding William Rogers, who was engaged in mapping a few thousand stars in the 'Harvard Zone' of the sky as part of a larger project launched by Germany's *Astronomische Gesellschaft*. Towards the end of her life, when she was in truly desperate financial straits, Pickering approached Harvard about obtaining a pension for her in recognition of her work, and the efforts of her father and brother in founding the university's observational programme. Harvard refused, and so Pickering dipped into his own pocket in 1906 to create a pension fund for her, on the understanding that nobody would tell her the source of the money, as he knew that she would not accept anything that looked like charity.

Anna Winlock, meanwhile, was the daughter of the Observatory's third director, Joseph Winlock (director 1866–1875), who was Pickering's predecessor in the role. When her father died in 1875, the year of her graduation from high school, it fell to Winlock to provide for her mother and four siblings, and so she turned to Harvard for work. She knew that the suddenness of her father's death had left many of his observations of the last decade in a frightful state of incompletion. Harvard wanted that data organised and compiled, but did not want to pay anything like a decent salary to do it, and so Anna Winlock seemed like the solution to all of their problems – somebody intimately familiar with the work, who had been trained by Joseph, who was devoted to seeing it through, and who, as a woman, could be paid half as much as a man.

Winlock would continue working at Harvard for the rest of her life, spending twenty years working on the *Astronomische Gesellschaft* mapping project that Bond had also contributed to, and which ultimately resulted in a chart containing details of about a 100,000 stars. As part of this and other work, she was often involved in meridian circle computations, which were for many years astronomy's most reliable source of positional information about stars. Meridian circles were telescopes which were locked into a position that could scan along the meridian (the great circle that runs through the celestial north and south poles, and the local zenith). Rather than chasing after stars, these telescopes waited for stars to pass over them, and took down information about the timing of that procession and the angles to the observed star. Taking that information and changing it into stellar positions was an intense process that Winlock mastered and employed over the course of nearly three decades of work, while simultaneously overseeing the thirty-eight-volume Observatory Annals, which collected tables of variable star positions and observations within star clusters. Winlock died suddenly, in 1904, at the age of 46, leaving behind her sister Louisa (1860–1916), who had served as a Harvard computer since the early 1880s.

Mary Adela Blagg (1858–1944)

It is difficult to know which of Mary Blagg's accomplishments to highlight as her most salient contribution to astronomy. Personally, the maths nerd in me loves her 1913 correction of Bode's Law. That law, otherwise known as the Titus-Bode Law, had worked for some time as a good general rule of thumb for the positions of the planets – namely that each planet is roughly twice as far from the sun as the planet before it. This works very well for a number of planets – Venus is about 1.87 times further from the Sun than Mercury is, Saturn is 1.84 times further away than Jupiter, Uranus is a lovely 2.00 times further than Saturn. It is pretty clever! The problem is Neptune, which was nowhere near where the Titus-Bode Law said it should be. It is, on average, about 1.57 times further away from the Sun than Uranus, which is a miserable 78 per cent of the distance predicted by the law. Blagg used her mathematical abilities to perform a Fourier Analysis (which involves representing functions as an infinite series of trigonometric quantities) that corrected the law, and has since been upheld as an accurate predictor of satellite locations.

While that feat is my favourite tour de force by Blagg, most Blagg fans (whom I am proposing we call 'Blaggers') break down into two warring camps as to what they find most significant about her – those who love her work on variable stars, and those who herald her as the individual who straightened out the Moon, once and for all. As to the first, from 1914 to 1918, she did the lion's share of the work in publishing a series of ten roughly twenty-page reports on variable stars that involved complicated analysis of those stars' variation in brightness over time. One of the most startling results of her work, published in 1924 and further analysed in later papers, was the case of Beta Lyrae, a star which she found had a gradually lengthening cycle of brightening and dimming in some grand cosmic *ritardando*. The phenomenon, it turned out, was a real one, caused by one of the stars in Beta Lyrae essentially eating the other one, drawing off its gases in a grim act of stellar cannibalism.

In 1905, Blagg had been asked by the International Association of Academies to create a list of lunar features, a task which occupied her for eight years and made her realise the full extent of the jumble which lunar taxonomy was in. She published her first list of lunar formations in 1913, along with her observations of all the problems with lunar nomenclature as things stood. Over the following years, she was absorbed in her variable star research, and did not have a chance to return to the Moon until 1920, when she joined the International Astronomical Union's Lunar Commission. Together with amateur astronomer Karl Müller she was tasked with standardising the names of the lunar features, some of which had half a dozen different competing names and descriptions. That project occupied fifteen arduous years, and resulted at last in the definitive two-volume *Named Lunar Formations* of 1935, which served as the standard work on the topic until supplanted in 1966 by the *System of Lunar Craters*, which had the benefit of employing photography generated during the American-Soviet space race.

For all of this, and more besides, Mary Blagg was justly one of the five first women inducted into the Royal Astronomical Society, in 1916, and even more justly, currently has a 5km impact crater named after her on the lunar surface whose chaos she gave a decade and a half of her life to taming.

Florence Cushman (1860–1940)

Florence Cushman was a bedrock presence at the Harvard Observatory, serving as a human computer from 1888 to her retirement a near half a century later in 1937. She served in both the Pickering and Shapley eras, and was particularly associated with her work from 1918 to 1934 on the Henry Draper Catalogue project.

Dorothea Klumpke-Roberts (1861–1942)

Though she was born and died in San Francisco, the majority of Dorothea Klumpke-Roberts' career was carried out in France, where she played a pivotal role in one of that country's major end-of-century astronomical efforts. Her father was a prosperous realtor in post Gold Rush California, and had the wherewithal to send Dorothea and her four siblings to European schools for their education. Klumpke-Roberts accordingly went to France in 1877, eventually studying at the University of Paris, where she earned her bachelor's degree in 1886. That was the year that two major astronomical projects were launched: the Henry Draper project at the Harvard Observatory under Edward Pickering, and the Carte du Ciel project, an atlas of the night sky to the 14th magnitude proposed by Sir David Gill, and which was seized on with particular enthusiasm by the director of the Paris Observatory, Amédée Mouchez.

Klumpke-Roberts joined the Paris Observatory in 1887, just as the Carte du Ciel project was getting underway, and it was this project that would consume most of her professional efforts for the next fourteen years. At first, she worked at the task of determining star positions from masses of photographic plates, much as Ida Barney would do a few decades later at Yale. The quality of her work was quickly noted, however, and in 1891 she was promoted to the position of Director of the Bureau of Measurements, beating fifty other male applicants for the job. Two years later, in 1893, she made history again by earning her PhD at the University of Paris for her dissertation on the rings of Saturn. It was the first non-honorary advanced degree in astronomy awarded to a woman, and was accompanied by admission into the French Academy of Sciences as an officer, another first for a woman.

In 1896, while on a solar eclipse expedition to Norway, she met the astronomer Isaac Roberts, a specialist in nebular photography. He was a 67-year-old widower, she was a 35-year-old trailblazer, but their shared love of astronomical photography drew them together, and in 1901 they married. Klumpke-Roberts thereupon made a hard tack in her career, shifting her research interests towards her husband's chosen field. The astronomical power couple decided to carry out a joint project to photograph the fifty-two 'areas of nebulosity' discovered by William Herschel (1738–1822). Isaac died in 1904, leaving Klumpke-Roberts to complete the necessary photographs, analyse them, and collate William's extensive past notes on nebulae with their joint observations, ultimately publishing the results in 1929 as *The Isaac Roberts Atlas of 52 Regions, a Guide to William Herschel's Fields of Nebulosity*.

In 1934, she was elected a member of the French Legion of Honour, and shortly thereafter returned to San Francisco, where she established a number of endowments to fund continued nebular research and support young astronomers. The Dorothea Klumpke-Roberts Prize, given by the University of California, Berkeley, continues to be awarded to this day, for seniors showing exceptional ability in mathematics.

to a pittance, and of those few real working years, most of her time was given to the pet projects of Pickering, who was interested in compiling data, not theorising. So, her work on Cepheids languished, while she examined plates from a dozen different observatories to produce the important but mechanical data Pickering craved. She was a computer who could *think*, and if she never had the freedom to let her curiosity guide her work, at least her results were important enough to pave the way for her intellectual descendants to carve out a small measure of investigatory autonomy for themselves.

Once intense multinational parallax studies, wed to the results of Doppler-shifted elemental spectra, allowed the ascertaining of certain Cepheid distances, the universe opened up. Soon, we could map any galaxy we could see. We found the probable size of the universe, and therefore its probable age, and thus learned truly the humbling scope of our place in it. And at the start of it all was Henrietta Swan Leavitt, the deaf human computer who saw a correlation when she was supposed to just record data, and thereby gave us the measure of the cosmos.

Further Reading

We have so little actual documentation about Leavitt's life that any book is bound to be a bit slim. I like George Johnson's *Miss Leavitt's Stars*. It puts her in the context of the dizzying rush to determine the shape and nature of the universe in the early twentieth century, and works to dispel some of the rumours that more popular accountings of her life wove into existence from nothing. I have read it maybe four times, and each time, it is still a fun and enlightening romp.

Chapter 15

She Filled the Sky
Annie Jump Cannon, Iron Woman of Astronomy

Three hundred and fifty thousand stars classified.

It is one of astronomy's unbreakable and frankly not even approachable records. Seven hours a day, six days a week, for forty-four *years*, one woman bent herself to the task of creating an ultimate chart of the night sky, with each star classified not only by position, but by surface temperature and spectral signature. Hunched over a magnifying glass, she could categorise three stars a minute where others might take three minutes to categorise one star. She was astronomy's Iron Woman – Annie Jump Cannon.

For the first thirty-three years of her life, Cannon shared the fate of most highly educated and intelligent nineteenth-century American women: a drifting existence desperately lacking purpose. Her mother had raised her with a passion for the stars. The two climbed the attic ladder nightly to look up at the sky and identify the constellations there. She went to Wellesley College and, in spite of an illness that caused substantial hearing loss, she graduated valedictorian in 1884. She was smart, young and passionate about science and, as far as society was concerned, she was *done*. She drifted for the next nine years, travelling the world, taking pictures with her beloved Kamaret, writing articles about the places she had visited, and always haunted by the knowledge that, with her intellectual gifts, she could be so much more. She poured her frustration into her diary:

> I am sometimes very dissatisfied with my life here. I do want to accomplish something, so badly. There are so many things that I could do if I only had the money. And when I think that I might be teaching and making money, and still all the time improving myself, it makes me feel unhappy and as if I were not doing all that I can.

She Filled the Sky

When her mother died in 1893, Cannon took it as a sign to *do* something with her life. She got a job teaching at Wellesley and began graduate studies at Radcliffe, an institution with ties to the Harvard Observatory. Cannon became one of Pickering's computers, and her work lay in steady exactitude, which she managed by refining the classification system she inherited from Fleming into the system that all astronomers everywhere now use to quickly categorise and rank stars by. Synthesising the various over-complicated systems of her day, she created the OBAFGKM scale of stellar intensity and its subcategories (O stars are the hottest, M the coolest), which for many years was taught to male undergraduates with the not entirely enlightened mnemonic Oh Be A Fine Girl Kiss Me.

Tasked with the massive, seemingly impossible, task of classifying all the stars in the sky down to magnitude nine (sixteen times dimmer than the unaided human eye can see) according to her system, in her forty-four years at Harvard she not only catalogued to magnitude nine, but to magnitude eleven, amassing in the process 350,000 stars in an era before computerised stellar recognition was even a wisp of a dream. By visual inspection, she had to interpret the achingly faint smear of a star's spectrum that had been produced by telescopes equipped with a spectroscopic wedge. It took keenness of eye married to a total familiarity with spectral astronomy and stellar composition to even begin to make sense of those smudges, let alone to process them definitively at the breakneck pace of three a minute.

Her work was so universally respected by the astronomical community that she was able to use her success to open the door to other female astronomers, endowing the Annie J. Cannon Scholarship for women entering the field. She was made curator of astronomical photographs in 1911 but was refused a professorship on account of her gender. Likewise, she was nominated for the National Academy of Sciences but was curiously rejected on account of her deafness.

After forty-four years of unfathomably steady work and dedication, Cannon retired in 1940 at the age of 77, and inevitably continued working anyway. She was a dedicated suffragist, and a tireless proponent of women in science, always willing to donate time and hard-won money to promote women of talent. Her work ethic was unparalleled, but all who knew her remembered her most for her kindness, for turning the drudgery of astronomical sifting into a personal adventure. Cecilia Payne-Gaposchkin captured that essence when she said:

> To [her friends] she was not the great scientist ... she was
> a human being, and as such they loved her – they, and their

children and their grandchildren. Perhaps the greatest tribute that I can pay to her memory is to say that she was the happiest person I have ever known.

Annie Jump Cannon died in 1941. The lunar crater *Cannon* is named in her honour.

Further Reading

In the Footsteps of Columbus is a curious book that Cannon wrote before becoming an astronomer, a collection of photographs and essays taken and written during a trip to Spain. It is an interesting document about the dawn of amateur photography but, more importantly, in the edition annotated by Doug West, there is the only substantial stand-alone biography of Cannon we have appended at the end. Apart from that, you have to make do with guest appearances in other books – George Johnson's excellent *Miss Leavitt's Stars* and Dava Sobel's equally wonderful *The Glass Universe*.

Chapter 16

Hydrogen Rules the Universe
Cecilia Payne-Gaposchkin and the Composition of Stars

'You are young, and wrong. You must retract.'

When fresh-faced zeal confronts experience, it usually loses. Scientists who think they have solved everything on day one usually find that they have merely wandered into a seductive semblance of a solution, the first of many in the years to come, and firmly correcting them is part of the sacred duty of every senior researcher. But what happens when the young person is right, revolutionarily right, and repeatedly revolutionarily right, and just as repeatedly compelled to retract their work?

What happens is Cecilia Payne-Gaposchkin (1900–1979). On at least three separate occasions, she advanced astronomical theories on the basis of overwhelming data, only to back away from them under pressure from her superiors, her insights only being ultimately validated later (often much later), when male researchers came to them independently. In the meantime, she slogged away, overseeing the categorisation of over 3 million stars for a paltry wage in the face of persistent denial of personal advancement.

Born in England to a family of ancient lineage, she was a polymath genius from the first. Her father died when she was only 4, but impressed in her a deep love of classical music that translated into a passion for violin playing and orchestral conducting later in life. At school, she studied Latin and Greek, and set about teaching herself French and German, just four of the six (or seven if you count her forays into Icelandic) languages she would ultimately speak fluently. She wrote poetry, could recite long passages of Shakespearean drama from memory, and generally had all the makings of a prodigiously gifted humanities student, were it not for the fact that she was *even better* at science and mathematics.

She initially had her heart set on botany, but in college one day she happened to hear a lecture by Arthur Eddington about relativity, and something deep clicked. She rushed home and wrote the entire lecture down verbatim from memory (it sounds remarkable, but this was something many people could do in the days before we relinquished control of our memory to portable storage devices). There was something in the mathematical sureness of Eddington as against the messy particularities of botany that she found fundamentally compelling, and she switched her course of study accordingly.

Her interest lay in the application of the new discoveries of emerging quantum mechanics to the understanding of stars. This was the late nineteen-teens, when we still did not really know what stars were made of, nor did we have the facilities to investigate how the ionisation of atoms works at exceedingly high temperatures. From the depths of quantum assumptions, however, some theoretical models were starting to emerge, and one in particular would have a profound impact on Payne. In 1920, Indian astrophysicist Meghnad Saha developed the Saha Equation, which described how temperature and pressure would affect the ionisation of atoms.

When you raise the temperature of a gas, the particles start slamming into each other and knocking electrons about, creating gas ions of various charges. Saha's model made predictions about the frequency of all the different ions you would find in the resulting plasma. And that is exactly what Payne, who was interested in using the spectra of stars to determine their composition, needed to explain the seeming chaos of the existing spectral data.

But to do that she needed a research position, and there simply were none in England to be had for a woman. Everywhere she turned, the advice was the same, 'If you want to continue in astronomy, go to America, and especially to Harvard.' And that is what she did. Harvard was, at the time, the leading astronomical institution in America, if not the world, and, as we have seen, it was built almost entirely on the work of women. Here Henrietta Swan Leavitt had discovered the correlation between the luminosity and period of Cepheid stars. Here Annie Jump Cannon had personally categorised over a quarter of a million Southern Hemisphere stars, and Antonia Maury a similar number of Northern stars, their combined work creating the mammoth *Henry Draper Catalogue*. And it was here that Cecilia Payne would, in just under two years, write the paper considered by many to be the best graduate astronomical paper ever produced.

She arrived in 1923, took over Henrietta Swan Leavitt's empty desk, and dived into the task of making sense of stellar spectral data. And it was a mess. Each star seemed to have an entirely different story to tell, featuring absorption and emission lines (frequencies of light swallowed and emitted when electrons hop between energy levels) that defied all ready categorisation. The reigning assumption was that stars must have elemental compositions similar to those on Earth – why should they not, after all? If Earth was made from thrown off bits of the Sun during its formation, why should the two not have roughly similar amounts of elements? However, the spectral data seemed to tell a different story, which it was Payne's fate to unravel.

Applying Saha's equation, she was able to explain the variations, not as indicative of the presence of different elements but of different ionisation rates arising from differences in temperature. As such, she was able to use the spectral data to create temperature scales of the stars in question, which would have been useful enough on its own, but in analysing the elements that were present, she discovered that hydrogen was over a million times more abundant than predicted by the 'stars are like Earth' theory. It was a radical departure from the common wisdom of the day, and both Harlow Shapley, Pickering's successor as Harvard Observatory's director, and Henry Norris Russell, perhaps the most famous astronomer of his age, put pressure on her to downplay her findings.

She was young, facing down two of the pillars of the astronomical community, both of whom she revered as essentially walking gods, and she did as she was told. In the paper she published, she concluded by saying that her theory about the frequency of hydrogen was probably wrong, and substituted a pet explanation of Russell's as to the most likely source of the discrepancy. She had been absolutely right but had backed down before authority, and not until 1928, when Albrecht Unsoeld's calculations of relative elemental abundances backed up Payne's original work, would the astronomical community as a whole recognise the value of what she had done as a mere graduate student.

From stellar spectra, Payne moved to a study of the super-giants, and found discrepancies in the data that she believed to be explained by interstellar gas absorption. Light coming from a star, she thought, was hitting clouds of invisible gas on its way to Earth. When gas gets hit with light, it steals the frequencies of colour from that light which will allow the electrons of its atoms to jump up a level. The light that makes it through, therefore, is different from the light that went in, and so the colours we see are different from those that originally left the star. Payne argued passionately with Shapley for the existence of a significant interstellar gas effect, but it contradicted one of his own cherished theories, and he pressured her to drop it, so she did. She was, once again, absolutely correct.

There was even a third incident where she was sure that the Stark Effect, in which atomic lines split in the presence of an electromagnetic field, occurred in stellar spectra, and found evidence of it in the helium lines of super-hot stars. She took her results to Shapley. He told her to bury them, and she did. She did not publish her data, and so it is Otto Struve who we now think of as the discoverer of this phenomenon, and not Cecilia Payne.

The story just gets sadder. The director of the Observatory wanted Payne to undertake the task of organising and standardising all of the data that had been piling up since 1889, and so she had to abandon her work on spectra to undertake the task that had similarly swallowed Swan Leavitt's research ambitions during Pickering's directorship. She did phenomenal work, systematising upwards of 3 million stars and discovering troves of new variable stars which opened up whole new fields of study, and she did it while editing all of the Harvard Observatory's publications, and for pay that was a fraction that of her male counterparts. For years, she was not allowed to teach under her own name, because Harvard's president refused absolutely to allow women full faculty positions, and when the position of director opened up, as it did on several occasions during her fifty-plus years working there, she was repeatedly and automatically barred from the position.

Why did she not she go to a different university, one which was amenable to female professors? Europe was out for political reasons, and because their record of letting women work in astronomy was still far less advanced than Harvard's second classing of them. There were other universities in America she could have gone to, but none of them had the sheer treasure troves of data that Harvard did. She stayed because it was there that she could do the work she loved doing. It was emotionally difficult, and at times subsistence was just barely maintained (she had to pawn her beloved violin one month just to have enough money to eat), but she could not leave Harvard behind because nowhere else had the means of investigating the questions that interested her, and I suspect the university knew this full well, which was why they were in no hurry to bring her salary up to a respectable level.

Time passed. Payne spent her rest hours, such as they were, conducting the observatory orchestra, creating art from improbable objects, raising children, and cooking elaborate meals. In short, her leisure hours were busier than most people's working hours, and her working hours contain contributions to science numbering in the literal millions. Her first paper, after its conclusions had been verified, made her famous, and her subsequent work had a longevity not given to many astronomers. Her citation rate was constant throughout her life, which almost never happens, and dipped only slightly after her death.

She was, and will always be, the first person to announce to the world the stuff of the stars.

Further Reading

Payne wrote an autobiography at the very tail end of her life which ranks among the greats of scientific memoir writing. There is a excellent edition of it edited by her daughter, Katherine Haramundanis, published by Cambridge, containing four introductory essays, two of which are great, one of which could have benefited from some editing (perhaps because it is by the editor), and one which rambles charmingly along here and there without any particular notion of Whence. Altogether, it is a great portrait of early twentieth-century astronomy in one of the world's leading institutions, filled with portrayals of an astounding assortment of women who silently shaped our knowledge of the universe. More recently, Donovan Moore has published a stand-alone biography of Payne-Gaposchkin, *What Stars Are Made Of.*

Chapter 17

Before There Was Sagan
How Helen Sawyer Hogg Brought Astronomy to the People

Before, 'The cosmos is all there is, all there ever was, all there ever will be,' there was, 'The stars belong to everyone,' the watch-phrase of a woman who spent three decades in energising the wider world, through her newspaper columns and television appearances, with a profound curiosity about the stars above them. She was Helen Sawyer Hogg (1905–1993) and to read her words is to be captivated immediately and profoundly by her infectious enthusiasm for the stuff of the night sky.

She was born and educated in Massachusetts, having graduated a year early from high school and been selected personally by astronomical legend Annie Jump Cannon to receive a scholarship that allowed her to study at Radcliffe with Harvard Observatory director Harlow Shapley. Shapley and Sawyer were both passionate about globular clusters, spherical collections of stars held tightly together by gravity. Studying the variable stars (stars whose brightness waxes and wanes periodically) in those clusters stood a tantalising chance of answering big questions about the size and age of the galaxy, as well as its development.

Sawyer worked with Shapley throughout the late 1920s in classifying the known clusters according to the stellar concentrations at their core, developing in the process the Shapley-Sawyer Concentration Class scale. During her time at Radcliffe, she met and fell in love with Frank Hogg, a stellar spectroscopy student whom she went on to marry in 1930. Until Hogg's tragic death at the age of 46 in 1951, he and Sawyer shared a life of mutual intellectual respect that serves as a happy counterpoint to the tragic marital tales of Mileva Maric and Harriet Brooks.

Together, they moved to Canada in 1931 to study first at the Dominion Astrophysical Observatory's 72-inch reflecting telescope in Victoria, and after 1935, at the cutting-edge David Dunlap Observatory at the University of Toronto, where Sawyer would remain the rest of her career. During her first years as a professional astronomer, she had three children and, rather than giving up her career, she integrated them into her nightly routine. While she sat on her observation platform dozens of feet in the air, her husband monitored the observatory control mechanisms below, keeping a watch on the baby. When the child cried, Hogg would descend, feed her, and then get right back up on the platform to continue her observations.

It is one of science's most beautiful images of a research-oriented family working together, sharing duties and supporting each other's work.

The work itself required patience on the superhuman scale. To produce clear photographs, Sawyer had to remain on her perch, eye glued to the eyepiece for as much as an hour at a time in order to make minute adjustments on the fly to the slowly developing globular image. Blinking was allowed, so long as you did not make too much of a habit of it.

Over five decades, Sawyer produced 2,000 photographic plates of globular clusters, but with that her work was hardly begun, for then she had to bend herself to the task of analysing them, superimposing images over their negatives, and employing a blink microscope to attempt to identify the variable stars that held the key to the clusters' age and distance. She produced three master catalogues over the course of her career, with a fourth in the works at the time of her death, which not only published new statistical data on thousands of known globular variables, but added over 100 new variable stars she discovered herself.

Sawyer found that, of 130 known globular clusters, only 108 of them had any variable stars, and only eleven of those contained more than fifty variable stars, providing important clues about the evolution of the clusters. In addition, her findings supported the earlier work of Pieter Oosterhoff that variables in clusters were found in two varieties, one metal rich (Population I), the other metal poor (Population II). The relative abundance of heavier elements is an important measure of a star's age – the younger you are, the more heavy elements you tend to have. The differences in composition Sawyer uncovered contributed to disproving the reigning theory that all the stars in the galaxy were the same age. The static universe gave way to a vision of stars emerging, dying, and producing in their turn new stars, and Sawyer's meticulous analysis and cataloguing helped pave the way for that vision.

As significant as her work was to the astronomical community (so significant that in 1946 she was the first woman elected to the Royal Society of Canada in physical science), most of us know and love her from her thirty years writing the weekly column, *With the Stars* in the *Toronto Star*. That column began in 1951, the year Sawyer's husband died, and a full twenty-nine years before Carl Sagan's Cosmos brought popular astronomy into American living rooms. In that space, she talked not only about the exciting picture of the universe being unveiled every day by professional astronomers but the breathtaking phenomena available to us all.

In 1976, she reworked her columns into a beautiful book, *The Stars Belong to Everyone*, which begins with sentiments so charming and inspiring that I have really no choice but to quote them at length:

> The first day in my new post I was interviewed by a reporter who asked brightly, 'Have you brought your telescope to Washington with you?' I had to confess I had not. 'My telescope' is a little job weighing 40 tons, with a revolving shelter weighing 80 tons. But after the reporter left, I realised that I had brought to Washington my capacity to enjoy the heavens. The beautiful phenomena attending moonrise and moonset and sunrise and sunset, the annual showers of shooting stars, Venus as evening star, the old moon in the new moon's arms – all these and many more were just as enjoyable and just as attainable for me as they would be if a giant telescope were standing by my side.
>
> Many people tend to postpone their enjoyment of the stars because they are constantly with us, but the iridescence of the twilight bow or the orange harvest moon rising slowly over a smoky fall landscape are celestial real-life scenes to look forward to from night to night or year to year. Very little time is required to see and enjoy the beauties of the sky; once you come to know them, they never lose their appeal.

If you are not, after reading those lines, compelled to go out tonight and put your back to the setting sun to enjoy the phenomenon of the twilight bow, then you are truly a lost soul and nothing can save you.

Sawyer also communicated her love of the stars on television, where, an irresistibly eloquent and passionate grandmotherly figure, Canadians fell in love with her all over again. Her degrees and honours are too extensive to list, as are the number of people working in astronomy today who owe their early inspiration to Sawyer's writings. With 200 papers of hard science, and thirty years of popular journalism under her belt, she set a standard by which to measure Sagan, Tyson and all who might come after, discovering the sky's secrets and taking the rest of us along for the ride.

Further Reading

You must find a copy of *The Stars Belong to Everyone*, buy it, and read it. Has astronomy moved on since 1976? Definitely. Will that keep you from being changed and energised from reading this book? Certainly not. Finding books about Sawyer Hogg is a trickier proposition. The great women in science pioneer Edna Yost dedicates a nice section to her in *Women of Modern Science* (1966), but Sawyer lived and worked a full two decades after that. To complete the story, head to the Astronomical Society of the Public's lovely obituary of her, written by Judith L. Pipher.

Chapter 18

She Followed the Sun
Ruby Payne-Scott, the World's First Woman Radio Astronomer

In March of 1944, a physicist and radar specialist named Ruby Payne-Scott aimed her equipment at the sky and became the world's first woman radio astronomer.

Within four years, she had established herself as a chief engineer of Australia's radio interferometer revolution, and one of the world's truly pioneering solar astronomers.

Three years later, her career was over. The woman who had contributed her engineering genius to the Australian war effort, and her keen mathematical sense to the nascent field of radio astronomy, was demoted to temporary status, her pension revoked, and the interest on her retirement payments reclaimed, when Payne-Scott's dark secret was uncovered by a routine departmental survey: Ruby Payne-Scott was *married*.

In fact, she had been since 1944, a fact which should have automatically disqualified her for a permanent position in the Australian government but which her colleagues, valuing her contributions and unique mind, banded together to help her conceal from the bureaucracy for seven years. Payne-Scott wrung every observation she could out of that time, acting as equal parts technician, astronomer, mathematician, inventor, physicist and engineer, hefting Australian science to its first taste of international prominence in the process.

She was 'forthright and outspoken' with a chronic inability to back down from a confrontation when she knew herself in the right. Dubbed 'Red Ruby' by her colleagues for her Communist sympathies (at least prior to 1956), she was a forward-thinking, shorts-wearing, equal pay-advocating atheist in an era that wanted nothing more than a return to a synthetic normality cobbled together out of Cold War fear and Atomic Era prosperity.

Born in 1912 in South Grafton, by 1915 her family had moved to Sydney and a world of expanded opportunities for an academically gifted young girl. Her mother, a former teacher, seems to have home-schooled Ruby and her brother through the elementary level; in 1923 Ruby commenced her formal schooling at Cleveland Street High School, where she graduated by the age of 13 with highest marks in mathematics and biology. Her star on the rise, she next attended Sydney Girls' High School, a highly regarded institution with a university prep focus. Again, she graduated early, leaving the school at 16 with first-class honours in mathematics and biology, and A's in every subject except French.

By 17, she was attending the University of Sydney with two merit scholarships supporting her studies, and her streak of academic excellence only continued as she completed the Honours Physics track in four years with another first-class honours and a string of scholarships and awards along the way. She was only the third woman in the university's history to take a degree in physics.

She continued on at the University of Sydney for her master's where, to gain money and experience, she brought her growing expertise in physics and mathematics to the university's Cancer Research Committee (CRC). Her work on the scattering of radiation in water was thorough, solid research that stood out favourably amongst the CRC's mainline efforts to validate the increasingly discredited theories of the magnificently monikered Dr Wanford Moppett. Payne-Scott earned her master's but a further career at the crumbling CRC was out of the question to anyone able to read the academic tea leaves.

And so, with characteristic energy and resolve, Payne-Scott threw herself into earning yet another degree, a Diploma of Education, which would allow her to take up a career in teaching should no new work in physics prove forthcoming. She earned that degree in 1938 and took up a position immediately as science mistress at Woodlands Church of England Girls' Grammar School. She was not, however, long for the world of teaching. Students described her as quiet and dedicated, but neither then nor later when she returned to teaching after being cast out of astronomy would her teaching career provide the satisfaction of her research work.

By 1939, Payne-Scott had left Woodlands and joined Amalgamated Wireless Australia as first a librarian, then head of the measurements' laboratory and finally, on the strength of her physics and mathematical background, as a researcher who soon displayed a gift for practical electrical engineering that spawned two published papers in 1941.

Then came the Second World War.

In 1941, Australia found itself suddenly at the front lines of a massive aerial and naval contest bloodily waged between the United States of

America and the Japanese Empire. One of the great strengths of the Allies in that conflict was the advanced state of their radar detection technology. Having joined the war, Australia threw itself into erecting a string of radar defence stations and funding new research into increasing the precision of radar measurements.

Upon finding that women not only could perform the work of radar operators but in fact routinely outperformed men in that role, the Australian government actively encouraged the training of women operators. Ultimately, 599 women worked as radar operators, a number hindered only by bureaucratic unwillingness to post women in forward combat zones. With her background in physics, however, Payne-Scott was destined to contribute to the Australian radar effort on a more fundamental level as a researcher, theorist and engineer. In 1941, she responded to a call for applications from the Radiophysics Laboratory (RPL) of the Council for Scientific & Industrial Research (CSIR).

The RPL boasted a core of engineers but stood in desperate need of a physicist with good mathematical abilities, so when Payne-Scott's late application came in with its list of published research from her time at the CRC and AWA she was quickly approved to join the team. For the next three years, she devoted herself to the thorny problems of radar calibration and the pioneering of interferometry to pinpoint hostile plane positions.

This latter work Payne-Scott would apply on a cosmic scale during her solar work of the late 1940s when she harnessed interferometry as the workhorse of radio astronomy. During the Second World War, this method centred on the use of cliff-side radar stations located along the coast. Ordinarily, bouncing radar waves off an incoming plane gave only a rough notion of the plane's location. But, by sending out a pulse and then measuring the interference of the directly returning wave from the plane and the wave returning after bouncing off the surface of the ocean, operators could radically improve the accuracy of their measurements. In perfecting these techniques and mastering the myriad factors that go into the proper calibration of radar equipment and the interpretation of incoming data, Payne-Scott developed precisely the skills that made her Australia's most valued solar observer for half a decade following the war's end.

The year 1944 proved fateful for Ruby. In March of that year, she and RPL chief J.L. Pawsey turned their 10cm receivers on the sky and so began the RPL's recorded climb as a respected centre of a research field that as yet had no name. Those measurements were largely left to lie fallow until after the war, but Payne-Scott's other great leap of 1944 would bring with it the potential for immediate consequences: her marriage in September to Bill Hall.

At the time, Australian law declared that women had to be demoted to temporary status, surrendering all pension benefits upon marrying; upon becoming pregnant they had to resign their position entirely. By 1944, Payne-Scott had fought hard for a permanent position with regular promotions and an impressive string of salary increases in recognition of her skill, and she was not about to surrender that position because of a law she considered medieval in principle and draconian in practice. Though officially registered as married, and though she informed her co-workers of that fact, there was an unspoken agreement at RPL not to inform the management in the bureaucracy of her new status. This act of necessary deception bought Payne-Scott a full six years of time to work and build and she used the time well.

At war's end, RPL faced a dilemma – they possessed a concentration of cutting-edge radar equipment and a staff of highly specialised engineers and physicists, but what to do with them? An extension of their earlier astronomical observations seemed a logical next step except for the small fact that not one of RPL's researchers was trained in astronomy. Payne-Scott saw the challenge before her and, characteristically, formulated a plan for how to subdue it, teaching herself solar astronomy and working out the adjustments to current radar technology that would be needed to make radio receivers a truly effective tool in solar observation.

In particular, she needed to tackle radar's imprecision when used on a galactic scale. Since radio waves have such massive, sprawling wavelengths compared to visible light, getting precise locations for objects as far away as the Sun was a dicey proposition unless new techniques could be developed. Wartime radar receivers, when pointed at the Sun, picked up a rich potpourri of radiation signals, but they had almost no ability to locate where on the Sun the signals originated.

Employing cliff-side interferometry methods, Payne-Scott was able to radically improve signal localisation, discovering that one particular type of polarised signal tended to occur in areas of sunspot activity (we now refer to this as a Type I signal). But cliff-side interferometers were drastically limited in their operating time, good for basically an hour of solar observation at dawn, and then lying useless the rest of the day. To create an instrument that employed interferometer methods with even greater accuracy and longer observational periods, Payne-Scott designed a 'swept-lobe' interferometer to be located at Potts Hill reservoir.

It was the world's first interferometer that swept radio signals across a surface at a rate of dozens of sweeps per second to capture solar detail, and was the first to use a movie camera to record the incoming signals. It would prove to be a mighty tool in twentieth-century astronomy.

While Potts Hill was under construction, Payne-Scott worked at a different observation site to which she had essentially been exiled after a clash with a fellow, male, researcher was decided in his favour – in spite of the fact that she was his senior at RPL. Working with one assistant, she set herself the task of validating a phenomenon nobody could quite bring themselves to believe. This involved what we now call Type III solar signals, which she essentially discovered and elucidated over the course of 1946. Unlike Type I signals, which appear simultaneously across all frequencies, Type III signals first appear in the upper frequencies, then make their way through the lower ones, with up to nine seconds of delay between their first arrival at a high frequency and their first detection at a lower one. Payne-Scott's work confirmed the reality of the frequency time lag and determined the unpolarised nature of Type III signals.

Following a miscarriage that caused her to leave work for six months, Payne-Scott found herself in 1948 the head of a new interferometry project, one of the best-paid scientists at RPL, and a researcher whose detailing of solar radio burst types was laying the groundwork for a new scientific discipline. She teased out the different strands of radio noise streaming from the Sun to determine how, based purely on radio characteristics, a solar flare could be distinguished from a sunspot or an ultra-fast electron stream pushing through plasma. She was at the height of her influence and inventiveness, when suddenly a bureaucratic marital status survey in the wake of a departmental rebranding brought to light her marriage to Bill Hall.

She was swiftly demoted to Temporary Status and lost all the RPL contributions that had thus far been made to her pension. She fought against the law as a gross and unfair act better fitted for Victorian England than Space Age Australia, but there was nothing to be done. A year later, when she discovered she was pregnant, she resigned her post, never to carry out original research again in the three decades remaining to her.

She threw herself into her new family life, raising two children, a boy who followed in her footsteps to become a mathematician, and a girl, Fiona Hall, who was for a time one of Australia's most beloved artists. During the 1960s, she returned to teaching, but between what students described as her harshly high standards and the creeping onset of Alzheimer's, the effort was not a resounding success. For the eleven years of her employment as a teacher, she never shared her time as Australia's leading solar astronomer with either her students or colleagues. Her memory failing and her classroom presence growing ever feebler, she retired at last in 1974, losing a little more of herself every year thereafter until, in 1981, she died in a nursing home in Mortdale.

Further Reading

Astronomer W.M. Goss has devoted a considerable amount of time to resurrecting the details of Ruby Payne-Scott's life and the nature of early Australian radio astronomy. His research is laid out in two books, 2010's *Under the Radar: The First Woman in Radio Astronomy* (co-written with Richard McGee), and 2013's *Making Waves: The Story of Ruby Payne-Scott: Australian Pioneer Radio Astronomer*. These are your go-to sources for work about her life, and have formed the basis of something of a Ruby Renaissance.

Chapter 19

One Life for the Sun
Hisako Koyama's Half Century of Solar Observation

If anybody embodies the spirit of Helen Sawyer Hogg's mantra that The Stars Are For Everybody, it is Hisako Koyama (1916–1997), who built a multi-decade career of profound astronomical significance without a university education or the mass of observational artillery we usually associate with twentieth-century astronomy. Instead, she found a subject she was passionate about, that she could study with generally accessible equipment, and stuck to her self-proclaimed task twice a day, every day, for nearly half a century, ultimately producing a body of data so uniform and reliable that it was chosen as her time's representative data set to stitch into a series of observations stretching backwards in time all the way to Galileo himself.

Koyama was born in 1916 into an urban Japanese family. At that time, thanks to educational reforms begun in the Meiji Era (1868–1912), education in Japan was in the midst of adopting Western models, with Tokyo Imperial University founded in 1877, followed in 1911 by the founding of the first co-educational college, Tohoku Imperial University, which opened its doors to women in 1913. Prior to that, women in Japan had no access to higher education, and in fact rarely were educated past middle school. In 1910, for example, of the roughly 3 million girls who attended primary school, only 56,000 went on to middle school, while of the 3.3 million boys in primary school, more than twice as many, or 122,000, went on to middle school. Fortunately, by Koyama's time, women's middle and high school education was significantly on the rise, and in 1935 there were actually more girls in middle school (412,000) than boys (341,000), though a significant part of that trend might have been more due to a curriculum change that taught girls sewing methods that were perceived by parents as useful in their future domestic lives, than a desire for women's academic advancement.

It was in the middle of this boom in girls' education that Koyama graduated from a Tokyo girls' high school, sometime in the 1930s. Her options at that point for continuing education were highly limited. She could have attended Tohoku, but it was located some 225 miles to the north-west of her family's home in Tokyo, which would have been an unthinkable distance for most families to send a daughter away by herself in that time. She could have taken the path pioneered by Tsuda Umeko, who went to the United States to attend Bryn Mawr college in 1899, but that was even more unlikely from a familial perspective than Tohoku. What might have been more acceptable was to attend the Tokyo Women's Normal School, established in 1875 (and still in existence as Ochanomizu University) as a training school for future teachers. The location was right, but the purpose was not. Though Koyama would later become one of Japan's leading lights for public outreach and education, she did not have any professed inclination at that time to become a teacher.

So, she rested content with her high school diploma, and threw herself into a private study of the thing that interested her the most: the stars. She bought and devoured books on astronomy, spent nights outside observing the sky with the help of astronomical charts and, after a visit to Tokyo's Tanichi Planetarium, resolved to make her own telescope, which she did with the help of an obliging telescope vendor, who showed her how to grind her own lenses. Later, her father purchased a 36mm refractor telescope for her, and it was this instrument that properly began her amateur astronomical career. There is a famous story of her, in the middle of the Tokyo blackouts during the end of the Second World War, dragging her futon out to the back yard so that she could sit under it with her star charts and take advantage of the sudden lack of city lights to properly observe the night sky, which is perhaps the most beautiful image of pure human curiosity that you will find in this entire book.

Her first desire was to make some useful observations of the Moon, but felt her telescope was not powerful enough to do any work of lasting value to astronomy; instead she turned it to the Sun, allowing the focused image to fall on to a sheet of paper, which she could then observe and sketch without danger to her eyes (the first solar telescopic observer, Galileo Galilei, famously did look directly through his instrument at the Sun on a few occasions, but took the precaution of doing so when the Sun was lower in the sky, and preferably in the presence of fog, to place as much atmosphere between him and it as possible, which was a *better* idea than looking at it at, say, noon, but still not a *great* idea.) In 1944, she anxiously sent her first sketch of solar sunspots to the Oriental Astronomical

Association's Solar section, unsure of whether her independent efforts would be lauded or condescendingly scorned. The answer she received from the section's president (and professor at Tokyo University), Issei Yamamoto, was positive, and for the next two years Koyama pressed on with her solar observations until, in 1946, she gained access to the 20cm refractor telescope at the Tokyo Science Museum (today the National Museum of Nature and Science) after joining its staff.

(For my fellow American readers who refuse to think in metric for moral reasons, 20cm is about 7.9 inches, meaning Koyama's telescope was significantly smaller in aperture than the 12-inch Whitin telescope donated to Wellesley for undergraduate use nearly three decades previously.)

This 20cm telescope would be Koyama's primary observational tool for the next forty-five years, until her full retirement in 1991. Every day, twice a day, for a period of about an hour each session, she would train her telescope on the Sun and draw the details of the sunspots she observed. Now, you might be wondering to yourself, with such an important long-term project, and with the vast improvements in instrumentation that took place over those four decades, why was it that she persisted in using that 20cm instrument from the war years? The reason lay precisely in the long-term nature of the project. To Koyama, and to the larger astronomical community, the continuity of instrument employed was one of the most important features of her work. By not upgrading her telescope every five years, she minimised the sources of variation in her observations, establishing a baseline of solar observation of unparalleled consistency.

Her first publication of her results in 1981, *35 Years with the 20cm Telescope*, and its follow-up volume, 1985's *Observations of Sunspots, 1947–1984*, were treasure troves of information for solar astronomers, and particularly in the field of sunspot study. In this, she was carrying on the work of Annie and William Maunder, whose research in the early twentieth century established the eleven-year solar cycle, and led them to create the Maunder Butterfly Diagram to visualise the distribution of sunspots over time. Koyama's work recorded three and a half of these cycles, which she, with characteristic modesty, would refer to as 'only a blink for the Sun' but that at the time represented the longest sustained set of sunspot observations in human history. That work has sparked even larger efforts to document the Sun's past, including a 2014 effort to construct a continuous sunspot history dating back to 1610 which employed data from Galileo, through Gassendi, Schwabe, Wolf and Maunder, and leaned heavily in the late twentieth century on Koyama. Further, her observations of sunspot asymmetries (also a speciality of Annie Maunder) were used by Derek Swinson in his studies of cosmic ray densities in the Sun.

Hisako Koyama was not just her era's greatest sunspot observer, however (though that would certainly have been enough accomplishment for one lifetime), but was also one of NMNS's most public representatives and advocates for amateur astronomy. When she was not recording her impressions of the solar surface, she was out on the museum grounds, organising regular telescope viewing parties for the general public, or planning planetarium shows to highlight interesting, readily observable features of the night sky, or writing popular articles to engage as many people as possible in the act of looking skywards and understanding what they are seeing. As an amateur astronomer who had been warmly encouraged on her way, she made sure that she gave of her time, encouraging the rising generation (and the one after that) to not give up on their passion for the cosmos just because they did not have a multimillion-dollar apparatus at their disposal.

In 1981, Koyama retired from the NMNS, but continued her research as a Fellow of the Museum until 1991, when finally, after forty-seven years of continuous observation, she said goodbye to our Sun at last, and headed off into the long sunset of her own life.

Chapter 20

Margaret Burbridge and the Dawn of Nucleosynthesis Theory

If you had asked a random astronomer in the 1930s how all of the elements in the universe were produced, they would have had a ready and instant answer: during the Big Bang event. Hypothesised in 1927 by Georges Lemaitre and heavily reinforced by Edwin Hubble's 1929 interpretation of stellar red-shifts as evidence of an expanding universe, the Big Bang seemed an ideal explanation of how the universe's elements arose – all produced at once in one grand moment of matter creation. That explanation, however, did not sit well with certain portions of the astronomical community, some of whom continued to hold to a 'Steady-State' model of the cosmos, in which the universe was eternal, and the elements were produced in a series of waves rather than all at once, and others of whom were simply ill at ease with the variation in element distributions across different stars that was difficult to explain by a single atomic origin event.

In 1946, the astronomer Fred Hoyle first proposed the idea, as a buttress to his Steady-State theory of the cosmos, that new elements could be created inside of stars, but that idea did not get much traction in the scientific community until Hoyle teamed up with William Fowler, Margaret Burbridge (1919–2020), and Geoffrey Burbridge to produce an exhaustive 100-page paper of such import and impact that it today is known simply as the B^2FH paper, which definitively laid out the modern theory of nucleosynthesis. The B^2 in that paper is a reference to the two Burbridges whose observations of stellar spectra and evolving theories about neutron capture as a motivator for the creation of new elements lay at the centre of the new theory, and whose path to the upper echelons of the astronomical community had been anything but a sure one.

Margaret was born Margaret Peachey (a family name originating, so Margaret believed, in Huguenot fisher, or 'pecheur', ancestors who came to England to escape religious persecution) nine months and one day after the signing of the armistice ending the First World War, a subject of much speculation on Margaret's part as to the circumstances of her conception, which her mother very properly declined to confirm or deny. Both of Margaret's parents were chemists (in fact her mother had met her father when she was taking a chemistry course of which he, some 17 years her elder, was the teacher), and in particular her mother had to fight for her education against her family's wishes that she settle for a more traditional trajectory. As such, it is not surprising to learn that young Margaret had access to essentially whatever scientific material piqued her curiosity. She was, from a young age, fascinated with abnormally large numbers, and used her father's binoculars, guided by advice in the periodical *The Children's Newspaper*, for which her parents bought her a subscription, to scan the night sky for objects of interest.

She was bought in turn a microscope, a chemistry set and, at around the age of 12, she was provided with the writings of a distant maternal relative, James Jeans (1877–1946), who was both a distinguished astronomer and an author of popular books on astronomy. It is probable that the book pressed into Margaret's hands was his 1929 *The Universe Around Us*, which excited her with its accounts of the unimaginably great distances between the stars. Years later, she recalled the thrill she felt reading that the nearest star was 26 trillion miles away – suddenly, her fascination with huge numbers clicked in with actual physical objects to which those numbers could meaningfully be applied. She formed at that moment a youthful conviction that some day, she would make the calculation of interstellar distances her profession.

While most of her friends at the Francis Holland School for Girls were interested in working towards attending Oxford, Margaret, on the advice of her mother, was destined for University College, London, where she matriculated in 1936. It was a good fit – UCL provided a rigorous slate of undergraduate science courses, and offered the option of a major in astronomy with a minor in mathematics that was tailor-made to Margaret's strengths and interests. At UCL, she learned all the practical skills necessary for making astronomical observations, maintaining and adjusting the instruments, determining errors in calibration, and working from raw data to orbital computation that would stand her in good stead not only in her first post-college job but in the variety of astronomical work that she performed across the long stretch of her diverse career.

Graduating in 1939 under the decidedly non-festive pall of coming war with Germany, Margaret first occupied herself as an air-raid warden, preparing London for the expected all-out air attack that failed to materialise during the period of the Second World War known as the 'Phoney War'. She soon found work more suited to her talents when the University of London Observatory's director and primary technician were both called up for war service. Somebody needed to be found with the technical skills to maintain the observatory's equipment, organise repairs in case of a bomb strike, and continue research with the 24-inch Wilson reflector telescope (the lenses on the ULO's refractor were deemed too valuable to risk keeping in operation during wartime, and were subsequently removed and separately stored). Margaret, with three years of technical training under her belt, and with no chance of being called up to serve in the war, seemed an ideal person to take on the job, and for five years she had constant access to a telescope, and the in-hindsight wonderful experience of carrying out astronomy in an open observatory, physically next to the device she was operating, something which she lamented later generations, huddled off in front of monitors stuffed into rooms far from the telescopes they controlled, rarely got to experience.

Following the war, Margaret attended graduate-level classes at UCL beginning in 1947, where she met fellow student Geoffrey Burbridge, whose interest at the time was physics rather than astronomy. The two were married in 1948, and thereby began their long career navigating the complex reality awaiting academic couples who worked in related fields. After their marriage, Geoff (as he was usually known) switched his field to astronomy, and together they decided to pursue a career in spectral analysis, which would require them to find a better telescope, in a less perpetually cloud bedecked city, than they had available to them in London. They applied for money to study at the Haute Provence Observatory, but were told that, if they wanted British money to look into a telescope, it needed to be a British one. So, the couple decided to self-finance their trip to Haute Provence, where they arrived in 1949 and fatefully happened to meet another visiting British astronomer, Fred Hoyle.

The Burbridges next move was to take some advice given them by the astronomical living legend of his age, Otto Struve, to apply for fellowships to study at American observatories. Margaret had previously applied for a Carnegie Fellowship to study at Mount Wilson Observatory, only to be told that Mount Wilson on no account accepted women researchers at its facilities, but fortunately both she and Geoff managed to find financing that gave them access to the Yerkes Observatory attached to the University of Chicago, where Otto Struve had gathered a team that included future Hubble Space Telescope guiding force Nancy Grace Roman and future Nobel Prize winner Subrahmanyan Chandrasekhar. Once in the United States, the Burbridges took every opportunity they had to study at different observatories and learn from far-flung experts who between them were producing the pieces that would lead to a radically new view of How the Universe Got Its Stuff. From

Harvard's Cecilia Payne-Gaposchkin, Margaret learned about elemental abundances in different stars. From Chandrasekhar, the couple learned new ways of relating spectral data to the underlying atmospheric physics of the star in question. From Maria Goeppert-Mayer's (more about whom in the physics volume of this series) theories about nuclear stability and a possible polyneutron origin of the cosmos, they began thinking about the role neutrons might play in progressions of element formation that might happen within stellar cores.

All of these influences, combined with their own measurements of atypical elemental distribution in the spectra of a series of Apm stars they observed, and with their 1954 meeting of Willy Fowler, coalesced into the project that would consume their mid-1950s and change the face of astronomy. Together with Fowler and Hoyle, Margaret and Geoff worked on establishing eight stellar processes that were important in the creation of new elements (nucleosynthesis): Hydrogen burning; Helium burning, the alpha process that occurs when Helium nuclei fuse and which can produce carbon; Hoyle's e-process that explained the relative abundance of iron as against other heavy elements through thermal equilibrium processes; the s-process of slow neutron capture that took place in the cores of red giants; the r-process of rapid neutron capture that occurs during supernova events; the x-process governing lithium, boron, beryllium and deuterium; and the p-process which sought to explain the production of lower abundance heavy elements. The results of their work, and the detailing of these processes, were contained in their landmark 1957 paper, 'Synthesis of the Elements in Stars' or the B^{2}FH paper as it is more commonly known.

That paper was a sensation that placed Margaret and Geoff squarely at the centre of the astronomical community's attention, and revived for a short while the hopes of the Steady-State universe advocates (of which Margaret was one) that their model, of progressive and sustained creation over time as opposed to a single origin event, might prove correct in the end. (That hope, by the way, was all but crushed by the 1964 discovery of the cosmic microwave background (CMB), the existence of which had been predicted by the Big Bang Theory but which did not fit in well with a Steady-State model.) In the following years, Margaret devoted herself primarily to the fields of spiral galaxies and quasars. In 1962, Geoff and Margaret joined the faculty at UCSD (University of California San Diego), which did not have the nepotism restrictions on husbands and wives working together in the same department that had so complicated their careers up to that point (and which similarly attracted the Goeppert-Mayers). From here, she carried out studies on the rotational velocities and mass to light ratios of spiral galaxies (the type of work that would later lead Vera Rubin to hypothesise the existence of dark matter), and her spectral studies of quasars, whose extreme redshift (and therefore, by the Big Bang account under Hubble's interpretation, extreme distance from the Earth) posed a problem for her Steady-State view of the cosmos which she attempted to rectify in 1967 by proposing a model for redshift that did not rely on an expanding universe.

Over the next decades, Margaret was dragged into the realm of academic administration on a number of occasions, in spite of her determined protestations that she was not temperamentally suited for the work. She became the first woman to direct the Royal Greenwich Observatory but found the politicking so to her distaste that she left as soon as she reasonably could. She was also chosen for the role of President of the American Astronomical Society, and almost immediately ran into the political kerfuffle surrounding the Equal Rights Amendment whereby some sections of the AAS proposed that no meetings should be held in states that refused to ratify it, while others vehemently opposed the mixing of science and politics. In addition, she was placed in the anxious position of having to refuse the Annie Jump Cannon Award in 1971, because she felt that it was prejudicial to have awards that were only awarded to women, and tended towards the under-valuing of their work.

Though much of her later career was spent in the deep brambles of administrative politics, Margaret Burbridge did have one major last, though ill-fated, contribution, and that was to a project overseen in its early stages by her old Yerkes colleague Nancy Grace Roman: the Hubble Space Telescope. The original plan for the telescope involved the inclusion of four instrumental systems, one of which was the Faint Object Spectrograph. In recognition of her years of spectral analysis work, Margaret was asked by NASA to join the FOS team. It was an important instrument that managed some key observations in spite of some terrible luck, first the famous aberration of the main mirror that rendered a number of the smaller apertures on the device useless, and then issues of insufficient magnetic shielding that created smeared pixels and effectively removed the 115–150nm part of the spectrum the FOS hoped to survey, and which meant it was unable to detect the important Lyman-alpha line associated with hydrogen electrons dropping from the $n=2$ to $n=1$ energy level. Ultimately, the FOS was replaced in 1997 by the Space Telescope Imaging Spectrograph, which is still in operation.

A hundred years from now, however, I doubt if anybody mentioning the name Margaret Burbridge will do so to reference her time as the Greenwich Observatory director, to wonder about her extended devotion to the Steady-

State model, or to gripe about design flaws in the Faint Object Spectrograph. What they will be doing is talking about the miracle year, 1957, when she was a member of that legendary quartet who rewrote the rules for why we have the elements we do in the amounts we do, and thereby brought us all to the realisation that we are, fundamentally, the children of the stars.

Further Reading

Unfortunately, there is no full-length biographical treatment of Margaret Burbridge currently available. Your best source for her life is a thirty-seven-page memoir she wrote in 1994 for the *Annual Review of Astronomy and Astrophysics* and which, amazingly in the world of online periodicals, you can access for free. There was also a series of articles in astronomy and physics journals and popular science magazines celebrating her hundreth birthday in 2019 which contain the highlights without the spectral details.

Chapter 21

Beatrice Tinsley, the Birth of Galaxies, and the Ever-Expanding Universe

Dr Beatrice Tinsley lived on our resolutely turning planet for only four decades, but in that time she gifted us a vision of the cosmos as dynamic and expanding even in those places it appears most placidly serene. She plucked galaxies from the pantheon of static celestial objects and showed us, through a rigorous combination of chemistry, mathematics, physics, and state-of-the-art machine computation, how these creatures are probably born, how they age and interact with one another, and what their fate will ultimately be. She was the passionate nerve centre of the world's cosmological efforts, connecting everyone to everyone through the medium of her encyclopaedic memory for facts and people, and when she was gone, there was simply nobody to fill the void. We had Tinsley, and then we did not, and astronomy was a long time in recovering from the transition.

The career of Beatrice Muriel Hill (1941–1981) spanned fourteen short years, from 1967 to 1981, but so intense was her passion for astronomy, so keen her basic need to work and discover, that she managed to forge the towering reputation of an elder scientific statesman in that brief span. There are people who take time to slowly develop their gifts and become the scientist they were meant to be, and others who explode out of the gate, shouting their skill from the first moment they turn their brain to the tangles and wonders of the world, and Tinsley was decidedly the latter. She tried everything, and succeeded at whatever she tried. She was a classical violinist who played chamber and orchestral music, a prize winner in Latin and French, and a lover of mathematics who sat in the back of the class just so she would not be distracted while she taught herself more advanced mathematics than her classes could provide for her hungry brain.

The daughter of an ex-military man and aspiring priest who moved the family from England to New Zealand in pursuit of a parish when Beatrice was very young, her skills shone in every school she attended, culminating in her college work at Canterbury University in Christchurch. At the physics department there, she threw herself into learning how to use the new computer systems to perform complicated calculations, a skill that would prove useful when she turned it upon the massive problems of galactic evolution later. And she fell in love, with a physics student named Brian Tinsley, whom she married some months before her twenty-first birthday.

Together, they relocated to Texas, where Brian had an attractive offer from the Southwest Center for Advanced Studies in Dallas to carry out his work in spectroscopy, while she was allowed to transfer her scholarship there to carry on studying. Soon after arriving in 1963, however, she decided upon the University of Texas in Austin for her PhD work, flying out to Austin every Tuesday morning and returning by bus every Friday afternoon to spend the weekends at home with Brian. It was a punishing schedule but worth it to finally be free of the sense of academic stagnation that she felt in Dallas, where she was often treated as more of a faculty wife than a scientist in her own regard.

Her PhD thesis dealt with the evolution of galaxies, and proved to be the start of a revolution in that field. In the face of an academic community which held that galaxies could be treated as fixed objects in the night sky because they did not effectively change their composition or luminosity over time, Tinsley argued that galaxies were in fact dynamic objects, the colours of which could not necessarily be directly tied to their age, and the luminosity of which fades distinctly over time, meaning that any attempt to use them as 'standard candles' is doomed unless the astronomer applies crucial correcting factors to their computations.

It was one of those theses that announce to the academic world the arrival of a new force, and demonstrated the hallmarks for which Tinsley would come to be known – a deep familiarity with the existing literature, and a fundamental ability to find and evaluate factors that influence astronomical events that others had not previously taken into account. It was the ship that should have launched a startling new career, but it was very nearly both introductory volley and swan song, as in 1968 the Tinsleys, unable to conceive a child of their own, took the step of adopting, followed a couple of years later by the adoption of a second child. Beatrice decided to stay at home to be with the children while Brian's work took him around the world for the standard academic fare of seminars, colloquia and consultations.

Over the next five years, Beatrice gradually returned to her research in astronomy, but it was not until 1973 that she was fully engaged in her life's work again, meaning that the scant fourteen years she had as a professional was effectively more like a brief nine. She was made aware of the fact that Austin was unlikely to ever give her a tenured position or resources commensurate with her research needs, and so she went hunting for a new university where she could realise the full scope of her intellectual vision. That university, it turned out, was Yale, but choosing it came at a great cost.

Beatrice and Brian agreed to begin divorce proceedings in 1973, but that left the question of what was to become of the children. When Beatrice decided to move to Yale, the question became that much more keenly pressing. To take the children with her would be to uproot them from their home, friends and school, but to leave them behind would mean only the barest of contact with them sporadically, as vacations might allow. After some trial runs, she satisfied herself that the children could do well without her daily presence, and made the difficult decision to leave them in Brian's care as she headed for Yale, and her destiny.

She arrived there in 1974, and the next six years saw her as the beating heart of a new community seeking to understand the universe through a complicated combination of physics, chemistry, mathematics, computer modelling and thermodynamics. The astronomy community, which had been so inspired by her PhD work, rushed to consult her, sending her papers for critique, and inviting her to conferences the world over. She took responsibility for the shepherding of graduate students, and made sure that they connected with experienced astronomers whose work aligned with their interests. She collaborated with observatories and her colleagues to produce dozens of papers which considered the impact of galactic collisions on the evolution of galaxies, the luminosities of galaxies in their infancy and demise, the existence and pace of universal expansion, the evolving chemical composition of galaxies as their component stars synthesised new elements (including FORTRAN programmes that modelled 12 billion years of galactic development), and encyclopaedic evaluation and criticism of the reigning galactic models.

She was *the* person you talked to if you had a notion about galaxies and the evolving universe and wanted to know what obscure but crucial chemical or physical interaction your model might be leaving out. Mentor, researcher and colleague, her frantic pace and iron work ethic astounded those who collaborated with her, and given the four decades of mature mental labour that ought to have been her due, who knows what visions of the cosmos she might have provided us.

It was not, however, to be. In February of 1977 she wrote her father to let him know that the doctors had determined that a lump on her leg was due to skin cancer. Throughout the next four years, she would experience bouts of seeming complete recovery, and pushed herself to work until the end, but in spite of several operations the cancer eventually spread to her vital organs. When the cancer invaded her brain, she lost the use of her right hand but, determined not to surrender to circumstance, she taught herself how to write with her left in order to continue her correspondence. Her last letter to her father, written in large blocky print, said simply at its end, 'I think of you a whole lot, not only on Birthdays, and wish you strength and happiness in the coming days. I honestly do not think the length of life is important. Very much love, from Beatrice,' next to which she drew a small picture of a beetle, which had been her nickname as a child.

At her funeral, one of her poems, written when she knew her death was imminent, was read aloud:

> Let me be like Bach, creating fugues,
> Till suddenly the pen will move no more.
>
> Let all my themes within – of ancient light,
> Of origins, and change and human worth –
> Let all their melodies still intertwine,
> Evolve and merge with ever growing unity,
> Ever without fading,
> Ever without a final chord …
> Till suddenly my mind can hear no more.

Further Reading

Edward Hill, Beatrice's father, wrote a memoir of her life after her passing, *My Daughter Beatrice: A Personal Memoir of Dr Beatrice Tinsley, Astronomer* (1986) which gives wonderful glimpses of her early years, though the later years are primarily given over to less-compelling accounts of her travels and meetings. There is, however, an introductory appreciation of her life, and an obituary by her colleagues that lay out her more scientific accomplishments, so all in all it is a fine book to have.

Chapter 22

Legacy, Suspended
Vera Rubin and the Ongoing Saga of Dark Matter

When Vera Rubin died in 2016, for a brief moment the gaze of the Internet swung her way to notice her remarkable life and work, to say a number of things about them of which perhaps half were true, and then moved back to the placid comfort of its usual stalking grounds. That instant of online recognition left a wake of bubbling misconceptions it will be the work of a decade to pop, but perhaps here, in honour of what Rubin did and fought for, we can make a start at sifting out some of the more tantalising fictions, to get down to the real nature of her revolutionary contributions to astronomy.

The first half of Rubin's career was governed by restrictions that had no right existing anywhere in twentieth-century science, but particularly not in astronomy. By the time she entered college in the mid-1940s, astronomy had blossomed under the work of a dazzling array of women observers and researchers, including Caroline Herschel, Margaret Huggins, Henrietta Swan Leavitt, Annie Jump Cannon, and Cecilia Payne-Gaposchkin, and dozens of others besides. Between them, they had created massive catalogues of the night sky, determined the constitution of the stars, provided measuring sticks for the night sky, and refined methods of observation and categorisation that drove the twentieth century's great astronomical projects, all fundamental contributions that, by 1948, no astronomer could ignore.

And yet, Rubin could not get a PhD from Princeton because the programme was, unfathomably, still not open to women. When she married at the age of 19, it was expected as a matter of course that she would follow her husband and make do as best she could with the learning opportunities available near his work, meaning that she had to give up study at Harvard, the pre-eminent astronomical institution at the time, for Cornell and Georgetown, which were not known for their astronomy departments. (Though to be fair, Rubin's husband is in a minority of mid-twentieth-century male spouses I have studied who actually made a concerted effort to equitably split household and child-raising duties with their scientific wives.)

The story that is often repeated, however, that Rubin moved into measuring orbital velocities of stars only after she was muscled out of her earlier studies by senior researchers who did not want a woman in their 'territory', is, by Rubin's own account, a gross exaggeration. Her early work included determining rotation curves and galactic velocities and studying quasars, but she found that she did not enjoy working in a field crowded with other researchers:

> I didn't like working on problems that many other people were working on and where I was constantly being besieged with questions about the work. I wanted a problem that I could sit and do at my own pace, where I wouldn't be bothered. ... We studied quasars for a year or two and I found it personally very distasteful. I just didn't like the pressure of other astronomers calling and asking me if I had observed this and if I knew what the redshift was. I didn't get to a telescope very often and it meant that I either had to give out answers that I was uncertain of, or say I hadn't done it and somebody else would then go do it. I just decided that wasn't the way I wanted to do astronomy.

(Bright Galaxies and Dark Matters, p. 157–158)

The constant jostling and pressure of highly populated research topics was simply not her style, so she found an obscure topic that she could do slowly and thoroughly, mapping the rotational velocities of stars in spiral galaxies using new instrumentation designed by her colleague Kent Ford. It was supposed to be useful, solid work with no surprises but, on the very first day, unexpected data started rolling in. The expectation was that, as with planets rotating around a sun, stars rotating around a galactic centre should show decreasing orbital velocities the further from that centre they were. At greater distances, the force of gravity should be less, and so, correspondingly, should the velocity decrease.

What Rubin and Ford found, however, was that the velocities flattened out, resolutely refusing to decrease with distance, a result that makes sense only if the rules of gravity work much differently than we believe, or if there is a lot more matter in those galaxies than we can observe through telescopes – dark matter.

Now, to be clear, Vera Rubin did not discover dark matter, and she was not the first person to note anomalies that implied the existence of unobservable matter. At least three different astronomers in the 1930s had made similar observations and one of them, Fritz Zwicky, had hypothesised that the discrepancies he observed were due to an unobservable form of matter which he dubbed 'dark matter'. He even came up with the theory of using gravitational lensing to further measure the amount of dark matter in the universe, a technique we use readily today but that was beyond the capacities of his instrumentation to measure.

So, no, Rubin did not invent the idea of dark matter or make the first observations that pointed to its existence. What she *did* do was produce the first modern evidence of it that made the astronomical community sit up and take the notion seriously. Her work was four decades after Zwicky's, and in the interim astronomy had more or less forgotten about his dark matter gambit. Rubin's velocity curves from the 1970s made the dark matter hypothesis impossible to ignore any longer, and the past four decades have seen an explosion in speculation about its existence and nature, from dark matter sceptics who believe that the observed phenomena can be explained by alternate means, to passionate advocates like Harvard cosmologist Lisa Randall who recently even used dark matter to explain the extinction of the dinosaurs (and it is not nearly as crazy as it sounds).

Rubin's ideas were lent support swiftly by measurements from other fields of astronomy, and in the popular imagination by the steady progress of neutrino physics, which verified the existence of curiously behaved particles that interacted only very rarely with matter. If billions of neutrinos can pass through you and not interact with your atoms, why can there not be a whole class of matter that will not interact at all with electromagnetic radiation?

And yet, dark matter still possesses a murky status. There are an astounding number of observations that suggest its existence, but there is also a dedicated body of physicists providing alternative explanations for those observations, such as Modified Newtonian Dynamics (or MOND), which Rubin herself spoke favourably of as a possible source for her measurements. There are experiments currently underway to provide final and indisputable evidence for dark matter's existence, but until that data starts rolling in, everybody invested in dark matter is playing a massive waiting game, with either Nobel Prizes waiting at the end or a frustrated but sportsmanlike shrug.

But for Vera Rubin the waiting is over; the clock has run out. The murkiness that characterised how best to interpret her revolutionary results

Right: Cheomseongdae is Asia's oldest surviving astronomical observatory. It was built in the mid 7th century by Queen Seondeok of Korea. (Wikimedia Commons)

Below left: Published in a small print run in 1650, *Urania Propitia* was astronomer Maria Cunitz's masterpiece, updating and simplifying the astronomical tables of Johannes Kepler. (Wikimedia Commons)

Below right: 1896 lithograph depicting Caroline and William Herschel. In a career spanning a half century, Caroline Herschel discovered 8 comets and 2500 nebulae. (Wikimedia Commons)

Jerrie Cobb was the leading figure in the Mercury 13, an early attempt to allow women to become NASA astronauts. (NASA)

On June 16, 1963, Valentina Tereshkova became the first woman in space. (Wikimedia Commons)

has spilt over into how to define her legacy. We have become so wrapped up in whether or not dark matter is a thing that we have overlooked the fact that Rubin's reputation ought not be dependent upon dark matter's existence or non-existence. She used new technology to study and analyse hundreds of galaxies to their very edges at a time when most astronomers were only interested in using it to peer deeper into the heart of the universe itself, and in the process produced the data that puts her orbital velocity model beyond question. And that model has created so many puzzles and so many theories and inspired so many people to enter astronomy that it is almost superfluous to ask which the right explanation will be.

So, that is Vera Rubin. She knew how she wanted to do astronomy, and found a way to make that happen, and in the process, created a mystery that drove half a century of research and the technological developments required to support it. Hampered initially by anachronistic restraints (though she did manage to discover the supergalactic plane when she was but 22), she found her way eventually and even credited her odd education with her success in looking for interesting projects away from the spotlight. Unlike the matter she might have uncovered, she shone in the darkness, and will continue to do so as long as we wonder about what might lie in that great vastness we shall probably never see.

Further Reading

There is a collection of Rubin essays and interviews, *Bright Galaxies and Dark Matters* (1997) that contains some biographical material, along with various articles on dark matter, supernova physics, the historical development of astronomy, and the question of women's place in physics, all written with an engaging zest for finding things out. More recently, in 2021 Jacqueline and Simon Mitton wrote a stand-alone biography of Rubin, *Vera Rubin: A Life*.

Chapter 23

Of Listening and Waiting
Jill Tarter and the First Forty Years of SETI

For scientists engaged in speculative research, the invisible adversary is nothing less than science's own history of conspicuous success. For 500 years, we have been pointing at problems and watching as science regularly solves them. We thought flying would be cool. Science handled it. We believed polio was rather a bad thing. Science locked that down too. We wanted to be able to walk around and virtually capture imaginary combat animals. Science paused a while, wondering perhaps about our priorities, but did it anyway.

Our every technological whim indulged, we are impatient and sceptical of any project that refuses to forecast semi-immediate success. And where there is money involved, that scepticism shades to viciousness quickly.

Witness the Search for Extraterrestrial Intelligence, or SETI, the umbrella term for a dozen branches of scientific research that have as their goal the describing and finding of possible extraterrestrial life. Over the last half century, it has been pummelled by politicians and satirists, scorned by mainstream astronomers, and forced to eke its way through funding complications, tortuous even by scientific standards. And at the grinding centre of the movement for forty of those years was astronomer Jill Cornell Tarter (b. 1944).

Her story reads like a Scared Straight programme to be shown to prospective science students, an object lesson in how scientific idealism gets caught and crushed between the gears of political opportunism and inter-disciplinary jealousy. For all the frustration her future would hold, however, her early youth was a veritable idyll of tomboying about with her father, fishing and taking things apart and blissfully ignoring her mother's entreaties to take an interest in 'girlish' things.

But that story came to a hard stop when her father died when she was but 12 years old, a victim of cancer, as her mother would be decades later, and as she herself would grapple successfully with in mid-life. Suddenly, an earlier promise she had made to her father to become an engineer took on the hue of a life-directing vow and she bent herself to maths and the sciences, attending Cornell as the only woman engineering student in a class of 300.

She powered through the isolation, labelled a hopeless nerd by the other women students in her dormitory, while being excluded from the all-male study sessions populated with people as interested by science as she. For graduate school, she turned to astronomy, which was just entering its third Great Age. If the eighteenth century flowered under the direction of Newtonian Mechanics, and the late nineteenth under the application of stellar spectroscopy, the mid- to late twentieth, was awash with new and exciting objects revealed by exquisite new instrumentation and methodology – quasars, pulsating neutron stars, black holes, and some shimmering signs suggesting exobiological elements.

The staid and steady universe was misbehaving and a new crop of astronomers rose to the challenge of sussing out explanations for these new behaviours and phenomena. Tarter was drawn to this world of galactic-scale questions, and was starting to compile expertise in the programming of telescopic software routines, when a small cabal of scientists and thinkers began pondering a question – simple in its expression but massive and labyrinthine in its elaboration – *How likely is it that life has occurred elsewhere, and is there a way to discover it?*

There is a truly sprawling parade of scientific disciplines hidden under that question. Biological questions about what constitutes life, chemical questions of how life began and what it needs to thrive, psychological questions about the nature of intelligence, sociological questions about the rise and fall of civilisations, physical questions about how communication on a galactic scale might work, astronomical questions about where to look for communicative signals, and engineering questions about how to gather and filter the cacophony of electromagnetic galactic noise bombarding our planet. There is hardly a branch of science that SETI does not touch at its most basic level, and so, for a certain type of Big Thinker, there is hardly any field more alluring.

In 1975, four years after NASA sponsored an initial survey into the feasibility of a SETI-type programme (named Project Cyclops), Tarter approached SETI pioneer John Billingham, and volunteered her services, thereby making the first step in a journey that would encompass four decades of struggle, triumph, and deep frustration.

All of that struggle can be summed up in one word: FUNDING. When SETI was primarily funded by NASA, it was routinely the target of flamboyant senators who saw they could earn easy political points by publicly mocking SETI as a waste of public money. Just like when the new kid in school realises he can survive by finding the class nerd and rubbing his face in the dirt, so did a generation of politicians view criticism of SETI as a sure, if callow, way on to the evening news.

Tarter's team had to fight for every dollar to buy time at telescopes, develop new software and receiving equipment, and pay the staff. They worked under a succession of vague, non-SETI-sounding acronyms to keep the senatorial predators off their scent, while pushing ahead knowledge on a dozen different fronts. Working on shoestring budgets with the constant threat of complete shutdown looming over their heads, they somehow managed to piece together answers and policies to SETI's deep questions: What distinguishes a natural radio signal from an artificial one? Is it better to survey all the nearest stars or to sample a bit from every sector of space? How do you electronically coordinate an alternate telescope to confirm a likely signal from your primary?

By 1992, Tarter and her team had the strategy, the software, the equipment and the NASA funding to begin observations at the massive Arecibo Observatory. Experience should have taught them to keep their heads low and get their work done, but instead they chose to launch on the five hundreth anniversary of Columbus's landing with a press conference boldly announcing their existence and intentions.

Politicians who had assumed SETI long dead rose from their slumber as one and resumed their old refrain. And, just like that, after a decade and a half of planning and research, Tarter's programme was cut again. A lesser person might have taken this subtle hint from the universe and returned to the fold of traditional astronomy, but Tarter knew the questions she had posed were important, perhaps even fundamental to our basic sense of our own humanity. If government would not fund her, perhaps industry might.

She needed a concentrated cadre of science enthusiasts and earnest futurists who also happened to be grotesquely wealthy. In 1993, that meant one place: Silicon Valley. Bill Hewlett and Dave Packard of Hewlett & Packard, gave $1 million each. Microsoft's Paul Allen dug deeper still and soon $7.5 million was raised. Plans were drawn up for the Allen Telescope Array – 350 radio telescopes that would deliver unparalleled resolution, bristling with state-of-the-art electronics and next-wave processing power.

It was a brilliant vision, but it would require time to assemble and, the longer it took, the scarcer on the ground investors became, nobody wanting to be the last supporter of a doomed cause. Ultimately, only forty-two telescopes were constructed by 2007, while Tarter herself retired from her position at the SETI Institute in 2012 to free up funding for the payment of other scientists' salaries.

Still active as the world's most famous SETI advocate (she was the basis for Ellie Arroway in the novel and film *Contact* by Carl Sagan), her career at the helm of a deeply committed but often rogue scientific vessel serves as inspiration to idealists and a cautionary tale to careerists – *somebody* had to take the knocks, and weather the ridicule, to give birth to a science so vast it might not secure its end in a hundred years or a thousand. Somebody with the grit to survive but the imagination to plan for a future still seven steps away. Somebody who could take satisfaction in putting the pieces in place that her intellectual descendants might know the thrill of announcing to a divisive world: There Are Others.

The technology and knowledge developed in assembling the SETI programme have vast merit on their own, but it is the cause they were created to serve – Tarter's cause: our cause as reflective and curious living beings – that will live, beyond Tarter or you or I, to that distant fringe of a day when our planet becomes a bit more humble, and a bit less alone.

Further Reading

Making Contact: Jill Tarter and the Search for Extraterrestrial Intelligence (2017) by Sarah Scoles is the book, benefiting from hours of interviews with Tarter herself. The organisation of the chapters can be disorienting, and I would have liked more on Tarter the scientist, but as the first biography to give Tarter her full due, it is a necessary part of anybody's Women In Science library. Also read Carl Sagan's *Contact*, because it is wonderful.

Chapter 24

Nancy Grace Roman and the Birth of the Hubble Space Telescope

It is an image to inspire almost primordial awe, taken when a space telescope with a rocky past pointed itself towards a black expanse of seemingly empty space and slowly accumulated flecks of light over the course of a million seconds. Here on Earth, we can see nothing in that black, but to the Hubble Space Telescope, there awaited a symphony of creation to behold, a field of galaxies from the early universe crowding the image like so many paramecia in a petri dish. In that one picture there are hundreds of cousins to our Milky Way stuffed into an arc of sky less than a tenth the diameter of our Moon.

It is an image that gave us our first true visual sense of our small if comfortable place in the universe, and yet it is but one of a million such that have remade the face of astronomy since Hubble became truly and completely operational in 1993. And without Nancy Roman (1925–2018) it might never have happened.

'The Mother of the Hubble Space Telescope' or simply 'Old Mother Hubble', Roman was a keystone figure in the history of NASA. That organisation's first female executive, and its first ever Chief of Astronomy, she was the person responsible for Getting Science Done in an era dominated by the distracting heroics of the astronaut programme. While the media focused on the drama of the Mercury missions and the lunar landing, it was Roman who ground away at the task of determining what science astronomers most wanted to do, and finding ways to harness the power of NASA to accomplish those projects.

It was a hard, important job, with many trailbreaking successes to celebrate, and it did not end happily.

The beginnings of Roman's astronomical career, however, were thoroughly lovely. Her father was a geophysicist who loved answering her questions about the world around her, and at the age of 11 she organised her schoolfriends into a miniature astronomical society propelled by Roman's voracious appetite for astronomical information. She graduated from high school in three years and got a position as a research associate at the legendary Yerkes Observatory (where she met Margaret Burbridge) after graduating from the University of Chicago.

As interesting as her research there was, however, there was no way to turn her position into a tenured one, so she transferred to the Naval Research Laboratory in 1955 where, because the scientists there had had a bad experience with the last woman scientist they had hired, Roman was virtually ignored, given no project to work on and no real equipment to work with.

During her years as a researcher, she produced several noteworthy discoveries, including the fact that not all stars of the same composition are of the same age, and how star velocity varies according to elemental composition, but the great break in Roman's life came with Sputnik and the subsequent creation of NASA in 1958. The new agency needed somebody to coordinate and define its purely scientific programmes, and one of Roman's NRL colleagues approached her with an offer to become NASA's first Chief of Astronomy.

It sounds like a no-brainer – who in their right mind *would not* want to become the first Chief of Astronomy at NASA? But it was a decision of weighty portent, for managing NASA's science programme would mean giving up her own research, perhaps forever. As NRL, however, was not precisely falling over itself to support her research goals, she chose NASA and a path that would lead her to the creation of the Hubble Space Telescope.

Roman's job at NASA was, in essence, to communicate with the astronomical community about what projects would be the most useful to do scientifically, evaluate their proposals in the light of her own astronomical expertise and knowledge of NASA's capacities, and organise funding and staff for the projects she felt to be worthwhile. On her watch, NASA launched a small fleet of satellites for observing solar phenomena, measuring relativistic red shift, recording planetary phenomena, and measuring ultraviolet astronomical events that were inaccessible to ground observatories because of Earth's atmospheric layer.

She was a frank and cautious administrator, known for bluntly denying projects that she felt carried too much risk while not providing enough scientific pay-off, and her honesty earned her detractors among the astronomical community. It was a difficult position, caught between her former scientist colleagues who could not or would not understand why she did not just approve everything they wanted, and her new employers at NASA who, especially after the lunar landing, were working with ever tighter budgets thanks to a combative Congress and an indifferent nation.

Roman's satellite fleet was a great step into making outer space humanity's new platform for scientific inquiry, but its crown jewel would take decades to come to fruition. What we now know as the Hubble Space Telescope was intellectually born as far back as 1946, when Lyman Spitzer wrote a paper on the need for space-based telescopes if astronomy was to solve its deepest riddles. Observatories on the ground labour under a myriad of handicaps. Our atmosphere buffets incoming light, nudging it hither and thither so that what entered as clear images of stars become, by the time they reach a telescope, hazy blurs whose mysteries can only be tenuously grasped by experts. And that is just what happens to visible light. Ultraviolet rays are blocked by the ozone layer, and with them all the information they carry about the stars that produced them. Water vapour and carbon dioxide, meanwhile, absorb infrared radiation which contains crucial data about the expansion of the universe.

On Earth, the rich buffet of electromagnetic wavelengths offered up by stellar and cosmic processes is reduced to a scattered snack plate of varying delectability. But a space telescope could feast upon it all, giving answers about how stars form and die, how galaxies are made, what the early universe was like, and what the ultimate fate of it will be, that we have sought since our origins as an intellectually seeking species.

Spitzer made a strong case, but it was up to Roman to propound the idea to astronomers and engineers. In spite of its advantages, it was not an easy sell. Ground-based astronomers saw a space telescope as competition, and a potential drain on their funding, and resisted Roman's overtures, while engineers doubted the feasibility of such a delicate piece of equipment weathering the launch process and being controllable in space. How would you remotely aim it accurately, and keep it focused on a single point while it orbits the Earth? How would you store the images? How would you get those images back to Earth?

Roman, Spitzer and a small cadre of other true space telescope believers used all their powers of influence and slowly pushed the astronomical community into some semblance of agreement about the general desirability of an orbiting optical telescope. Which brought a whole new problem: now that the astronomers were down with the idea of a space telescope, they ALL had ideas about what it should be like, and once again it was Nancy Roman who had to stand in the crossfire and be the adult making tough decisions that bruised egos.

Negotiating between those who wanted giant optical telescopes and those who wanted a Swiss Army telescope bristling with lots of specialised (and often contradictory) pieces of measuring equipment, between NASA administrators who imposed budgetary restrictions that nearly killed the project and a Congress who could not understand why they should fund a space telescope when they *just built* a large array of radio telescopes on the ground at great expense, Roman kept the project alive long enough to build a dedicated team of researchers, engineers, lobbyists and administrators who overcame the technical challenges of space-based astronomy (including the development of CCD technology that became the basis of the digital revolution in photography) and placed tactical pressure on government to support further planning and design throughout the 1960s and early 1970s.

And that is when the story goes awry. Roman had been NASA's reigning advocate of the power of space astronomy for a decade and a half, had made its first satellites a reality, and had thrown together support and funding for the Hubble Space Telescope, but the mid-1970s saw her pushed increasingly from the source of the action. In 1974, a reorganisation of NASA placed her one step lower in the administrative hierarchy, with a new overseer who parcelled the work she used to do out to other people, even to people who technically ranked below her. Given less to do, and finding little support from the astronomical community that still rankled from her frank approach to project rejection, she retired from her post in 1980.

After NASA, she worked as a consultant into her seventies, watching from the sidelines as Hubble laboured its way through final funding and construction, was launched to acclaim, produced its first images in 1990, indicating a massive problem with spherical aberration that rendered it all but useless, before emerging from the ashes after a repair mission in 1993 to produce the images that entranced and inspired a generation. And it is still up there, measuring where our universe came from and where it is going and making the scale and objects of the cosmos tangibly real to elementary school students and laymen.

Roman could be forgiven if she was bitter about missing out on the culmination of her efforts, or about what she suffered as a woman administrator in a man's environment. But, somewhat superhumanly, she is not. Here's what Women@NASA head, Mamta Patel Nagaraja, had to say about her opportunity to talk with Nancy Roman:

> I've interviewed Nancy Roman myself informally in her apartment in Friendship Heights. It was an incredible experience – she was gracious to invite me over, prepare tea and biscuits for me, and just talk. Like I was as important as she was to the world of women in STEM. It was immediately apparent to me that she understood what it meant to be a role model and that something as small as an hour chat was enough to inspire a woman to continue in the world of engineering and science. I was so impressed with her poise and wisdom. I recall when I expressed frustration that we were STILL talking about women in STEM from when I first heard it as a teenager, she calmly said 'Mamta, change takes time. Generations.' That's when it clicked for me – this has been and will continue to be something we just have to invest in for the better of the future. Things are much better than they used to be so clearly change has happened but we just aren't there yet.

Here's hoping equity takes a bit less than generations to happen, and that when it does it will be because of a new vanguard of children who take a look at a picture containing a thousand galaxies and whose curiosity leads them to answer the question 'Who made that possible?' and find inspiration for themselves there.

Further Reading

If you want to know about the nightmare of getting NASA projects approved, funded and built in the 1960s through to the 1980s, Robert Zimmerman's *The Universe in a Mirror* (2008) tells the tale with merciless clarity as he unravels the creation of the Hubble Telescope, what went wrong, and the heroes behind its repair and continued operation. He has a number of pages there dedicated to Roman's life and work, as well as to those of half a dozen other people who gave a decade of their best years to Hubble only to be shoved aside and eventually forgotten.

Chapter 25

Studies in Expectation
Jocelyn Bell Burnell and the Discovery of Pulsars

Sometimes the hardest thing about living in the universe is knowing about it. There is real and true terror to be faced in the indifferent march of the universe towards its inevitable dark close, and it takes a special type of person to stare into the void of our cosmic destiny on a daily basis and contemplate the mechanics of the vast machinery, coolly plotting the complete demise of us, everything we have ever done, and everything we shall ever do. Psychologically, astronomers of the modern era have to be made of some pretty stern stuff, and how they cope with the magnitude of their subject matter is often as fascinating as their research, and few in this regard are as consistently compelling as Quaker astronomer Jocelyn Bell Burnell (b. 1943).

Bell Burnell has a gift for Taking Things In Stride that has served her well both in the contemplation of humanity's grim chances in a hostile universe and during a life that, while serving up an exciting set of triumphs, has also dealt her more than her fair share of restrictive expectations. Her equanimity in the face of adversity is perhaps the result of her unique upbringing – she was the eldest sibling of a father whose family had Quaker roots going back to the seventeenth century and a mother who had been denied a higher education at the hands of the Great Depression. They were both determined that all of their children would receive a quality education, and that they would be brought up in a tradition of critical thinking.

Bell was born in Ulster, where the reigning Quakerism was of a fundamentalist, Bible-first variety that was not in accord with her parents' more critical and inquisitive approach to religion and the world. Their Quakerism was more of the English streak, in which one's own experience of religious moments had more weight than textual sources and church traditions. It was a religion that left room for critical thought and adaptation to facts, and one that had historically placed an emphasis on gender equality, and as a result had produced, over the course of the nineteenth and early twentieth centuries, an impressive line-up of women scientists including Maria Mitchell, Kathleen Lonsdale, Sara Josephine Baker, and Graceanna Lewis.

Soon after the Second World War, Bell's father resumed his career as an architect, to which work he would often bring his eldest daughter as a companion and helpmate, and so from an early age she became used to the careful use of measuring instruments and to the mental habit of checking one's results against one's common sense, of thinking to oneself 'is this measurement reasonable?' before blindly jotting it down on the page. These were habits of mind that would serve her well on the road to discovery and would make her an international name, but they had to wait for their development until Bell was moved from the generally poor quality state school she had attended for most of her youth to The Mount, a Quaker boarding school in York where she received exposure to the more contemplative bent of English Quakerism, and encouragement in the form of an elderly physics teacher who had come out of retirement to give the girls at The Mount a solid scientific education – Henry Tillot.

It was in these years that Bell realised the extent of her abilities in science, that not only was she doing everything more easily and naturally than those around her, but that she seemed to have an ability to explain difficult concepts to her peers to help them through their coursework. Science seemed to click, and she felt a particular pull to astronomy after her father sought out a copy of astronomer/science fiction writer Fred Hoyle's 1955 classic *Frontiers of Astronomy*. Her father, as it happened, was a consulting architect at Armagh Observatory, where one of the astronomers, in talking to her about what it was like doing astronomical research, told her that one of the practical expectations of the job was putting in long nights at the telescope. This deeply worried Bell, who knew enough of herself to realise that she functioned best on a regular sleep schedule and had no gift for pushing herself far into the wee hours.

Fortunately, a few months later, Bell discovered the existence of the field of radio astronomy, which answered questions every bit as monumental as those of traditional optical astronomy, but that did not require the night sky for its observations. Reassured, she began attempting to piece together exactly how one became a radio astronomer, and wrote a letter to radio astronomy pioneer Bernard Lovell, who had founded the Jodrell Bank Observatory in 1945 to study cosmic ray phenomena.

Lovell's advice was simple – do physics. If you have a solid enough basis in physics, everything else can be picked up along the way. Since

Bell was not attending a high school whose purpose was to feed students into the rarefied realms of the Cambridge-Oxford axis, she set her sights on schools of less towering reputation but boasting solid physics programmes, and ultimately chose the University of Glasgow.

At Glasgow, there was a tradition that, whenever a woman student entered the lecture hall, all of the male students would stomp the ground and bang their desks while shouting catcalls until the woman student took her seat, a ritual repeated to one degree or another whenever she asked a question, and with particular vehemence if she ever outperformed her male colleagues in class exams. It was a hard enough atmosphere to bear up against when there were other women in the class to lean on, but more often than not, Bell was the only woman student in her physics classes, and had to stand alone against the ritualised abuse which her professors did nothing to try and stop.

In spite of the discouraging campus culture, Bell continued with her studies in physics and some chemistry, bypassing entirely the positional astronomy that was all Glasgow at the time had to offer by way of astronomical coursework. Her end goal, however, was still radio astronomy, and when Jodrell offered its first summer school programme for promising undergraduates in 1964, Bell immediately signed up and was accepted.

That programme gave her hands-on experience of what doing radio astronomy research was all about, and also brought her into the orbit of the astronomers blazing a trail in the big-dish approach to radio astronomy. They encouraged her to apply to Jodrell for her graduate work, but her application was never answered, officially because it got lost, though Bell believed the somewhat famous misogyny of Jodrell's head astronomer also probably played a role. There was nothing for it, then, but to apply to Cambridge on the slim chance she might be accepted. Fortunately, as a candidate from North Ireland, she came with a grant that was tied to her person rather than her department, and on the strength of that already-secured funding her late application was accepted, placing her fatefully under the supervision of Anthony Hewish.

Hewish was studying quasars at the time of Burnell's arrival in 1965. These objects, first discovered in the late 1950s, consist of supermassive black holes (sporting masses on the scale of millions to billions that of our own Sun) surrounded by clouds of gas that release titanic amounts of electromagnetic radiation as they fall into the black hole. Hewish's project for Bell was to construct a radio telescope which was then to be used to hunt for quasars, the light from which presents tell-tale irregularities (or 'scintillations') as it passes through our Sun's solar wind on the way to our radio telescopes. By looking at hundreds upon hundreds of feet of radio telescope observations, Bell could determine the patterns for how quasar scintillation varied with the density of the solar wind, and thereby produce a rough estimate of the angular diameter of the quasar.

It was a clever idea for using the response of the quasar's light to the interfering presence of solar wind to produce an estimate on the size of the quasar, and for two years Bell spent her mornings in the bitter cold constructing the 4-acre apparatus that would do the job. Amazingly, it worked perfectly from the first throw of the on switch, a testament both to the soundness of her design and the thoroughness of her cabling. The telescope produced 96 feet of observational paper a day, which Bell had to comb through in the search for, and cataloguing of, likely quasars.

On 28 November 1967, Bell noted what she described as a 'bit of scruff' – Irish slang for untidiness – in a section of recorder paper about an inch in length. It did not behave like the scintillating sources she had been tracking so far, but it also did not have the right period for the other forms of interference she knew about. It was a mystery that her brain carefully logged away, and the next time she saw the anomaly she remembered its first occurrence and brought the irregularity to Hewish as a potentially noteworthy inconsistency. What was needed was a better image of the event, taken at a higher speed setting that would show in more detail what was occurring.

When Bell found the signal again, the high-speed paper revealed a remarkably regular pulse structure occurring every one and a third seconds, which Hewish took to be a sign of a man-made origin, since only very small astronomical objects could produce pulses of that period; objects too small to be stars. When Hewish stopped by the observatory and saw the event for himself, however, he began to be convinced that there was something potentially significant there, and set in motion the process of checking the phenomenon in another telescope, to eliminate the possibility that it was an artefact of some mistake Bell had made in constructing the device.

The secondary telescope picked up Bell's phenomenon, which was dubbed the 'Little Green Man' for the possibility that it might be a broadcast from an alien civilisation and not a legitimate astronomical occurrence. When Bell discovered a second pulsar, and then a third, however, in completely different parts of the sky, the possibility of it being an artificially generated signal was all but eliminated, and the Cambridge astronomy department was set in motion to organise the announcement of the discovery. Bell was not included in the meetings to discuss that strategy, and she was actively prevented by Hewish from changing the topic of her thesis from her quasar observations, which were an integral part to the research he was doing, to the discovery of the pulsar.

Bell's name was listed second on the paper announcing the existence of pulsars which was ultimately sent to *Nature*, and she included her discovery as an appendix in her quasar thesis paper so it would be firmly lodged in the scientific record. As a third-year graduate student, Bell had made one of the most important astronomical discoveries of the twentieth century, and had she stayed at Cambridge there is no saying what role she might have played in the studying and ultimate explanation of the nature of pulsars (Thomas Gold would ultimately receive credit for the correct model of a pulsar as a rapidly rotating neutron star).

It was, however, at this moment, between the discovery of the second and third pulsars, that she became engaged to fellow Quaker Martin Burnell, a government official who expected as a matter of course that the demands of his job would always take precedence over the career of any woman he should marry. The couple waited for marriage until after Bell received her PhD, but the next two decades of Bell's career would be entirely determined by what city Burnell had decided to move to in order to advance his own career, leaving Bell to grab whatever positions she could in the immediate environs, often with a reduction in pay and prestige.

She first moved to the University of Southampton, where from 1968 to 1970 she performed studies of the ionosphere in Michael Rycroft's group that she did not much believe in and did not go anywhere much, causing her to switch to Southampton's gamma ray group, which involved using equipment attached to balloons to make observations of electromagnetic radiation in the 1MeV-10MeV range. Bell taught classes while carrying on her calibrations of the gamma ray equipment, and discovered that not only was she good at explaining things to students but that she rather enjoyed it.

In 1973, however, her husband found a new job, which necessitated a new move, and Bell managed to find a position in 1974 at the Mullard Space Science Laboratory, where she worked as a member of the technical support staff for the Ariel 5 X-Ray satellite. It was a good use of the skills dealing with high-energy electromagnetic radiation she had developed with Southampton's gamma group, but it was considered a technician position rather than a full-fledged research science post. The job, however, was rewarding, as it placed her at the centre of the remarkable flood of data that the Ariel 5 produced – data which she would go on to mine for her own research goals.

In 1974 also came the announcement that the discovery of pulsars was being rewarded with a Nobel Prize – to Anthony Hewitt and Marin Ryle,

Bell's superiors at Cambridge. Bell had been cut out of recognition for the discovery of the object which had been detected by machines she had built, the signs of which she had detected on one inch among thousands of feet of recording paper, and the astronomical origin of which she had confirmed with her discovery of second and third pulsars. It was a glaring and controversial omission which Bell took largely in her stride, to this day insisting that she felt it would have been improper to give a prize for work done as a graduate student, and that she was simply happy that something she contributed to allowed astronomy to get its first Nobel for physics.

With the birth of her child in 1973, Bell quickly realised that she was not made to be a round-the-clock, stay-at-home mother, and looked for opportunities to keep in the world of astronomy and education that she could explore from home, opportunities which came her way in the form of the editorship of *Observatory*, a prestigious astronomical journal that had existed since 1877, and in her participation in Britain's famous Open University programme, whereby adults can further their education with correspondence courses, supported by tutors in their area. From 1973 to 1987, Bell worked as a tutor in physics and astronomy for the Open University, while working concurrently at Mullard from 1974 to 1982, and found that she loved working with people so motivated to learn that they stuffed their precious leisure hours with studying while holding down regular jobs.

Burnell's quest for advancement necessitated another move in 1982, this time to Edinburgh where Bell managed the running of the James Clerk Maxwell telescope, which involved interesting international coordination and travel, but tended more towards management than research. In 1991, she became professor of physics at the Open University, a position she held for ten years, during which time she succeeded in re-introducing original research as a significant component of the university. In 1993, Burnell, in a move that will, by this point, surprise precisely nobody, decided to leave Bell for a younger, less complicated, spouse, which ultimately allowed Bell to step forward into the spotlight in a way she had not allowed herself while married to a spouse who bridled at her success and fame. She took up more speaking engagements and interviews, and in 2004 took up the presidency of the Royal Astronomical Society for two years, followed in 2008 by a two-year stint as the president of the Institute of Physics.

In 2013, she published *A Quaker Astronomer Reflects: Can a Scientist Also be Religious?*, a transcript of a James Backhouse Lecture she gave to an audience of Quakers that stands as one of the more remarkable reflections about the intersection between religion and astronomical research yet set down by a scientist. It is unrelenting in its portrayal of a doomed planet set in an equally doomed universe which is consuming its precious reserves of hydrogen on its way to a cold, dark, diffuse nothingness. Bell's Christianity has no room for creation mythologies or all-powerful superbeings, indeed for anything but the assuredness of destruction and the comfort of some godly presence that means us well but is rather powerless in the face of the basic truths of existence. It is an intriguing, terrifying document but ultimately, even for those of us who do not believe in godly beings of any sort, a comforting one. Now is ours, tomorrow is not, and that makes our todays just that much more meaningful in their beautiful ultimate meaninglessness.

Jocelyn Bell Burnell continues to study intriguing aspects of the deep night sky, and in 2018 won the Special Breakthrough Prize in fundamental physics, the proceeds for which she donated entirely to a fund to help women, refugees and minorities achieve research positions, making the journey for the next generation perhaps a bit less fraught with expectational encumbrances than hers had been.

Further Reading

A Quaker Astronomer Reflects is pretty easily obtainable through print-on-demand type services, and contains bits and pieces of Bell Burnell's story in addition to a gallop through all of the different ways life as we know it is going to inevitably end. For the most in-depth look into her career this side of the three-part BBC mini-series *Beautiful Minds,* her American Institute of Physics interview with David DeVorkin is pretty wonderful.

Chapter 26

Jane Luu and the Discovery of the Kuiper Belt

For half a century after the discovery of Pluto in 1930, the common wisdom was that it was the last, furthest member of our solar system, and that beyond lay only darkness and dust. Though Gerard Kuiper had hypothesised in 1951 that there should be a rich band of asteroid-like objects at the edge of our cosmic neighbourhood, over the subsequent decades all attempts to find anything out in the Trans-Plutonic realm yielded nothing. The pessimism about the existence of the Kuiper Belt had become so omnipresent by the 1980s that, when Jane Luu (b. 1963) and David Jewitt (b. 1958) requested time on telescopes sufficiently sophisticated to potentially find objects beyond the known solar system, they were often turned down, as the venture seemed to some in the astronomical community a waste of valuable and rare telescope time.

These setbacks, however, did not deter Luu. Indeed, as adversity goes, academic refusal of observational hours hardly compared to the hardships Luu had already endured. She was born in Saigon in 1963, in a Vietnam entering its second decade of armed conflict, and for the first twelve years of her life, war was a constant companion. Her father was an interpreter for the US armed forces who put his often-insubstantial pay in the service of educating his daughter and two sons. When Saigon fell on 30 April 1975, he knew that his years of aiding the United States would be held against him and his family, and told his children to pack one bag each of only absolute essentials, for they would be leaving the next day. Luu did as she was told, though she did manage to sneak her favourite set of coloured pencils into her bag.

The family made it to the United States, but at first found it hard going. Upon arrival, the family was shuffled between refugee camps and motel rooms until Luu's father left to find work in California while she and her siblings remained behind in scenic Paducah, Kentucky, a sleepy town of some 31,000 souls at the time – which has since contracted to just over 27,000 where the biggest employers are the local hospital and WalMart.

By 1976, the whole family had reunited in Ventura, California, where Luu quickly added English to the French and Vietnamese she already knew and emerged the valedictorian of her high school in 1980, with a scholarship in hand to help defray the expenses of attending her college of choice, Stanford University. At Stanford, she majored in physics, but in between her undergraduate and graduate years she happened to pick up a job at Pasadena's Jet Propulsion Laboratories, where every day she passed photos of the solar system taken by the Voyager missions that JPL had had such a prominent hand in directing. The work being done at JPL inspired her to study the remaining mysteries of the solar system, and by 1986 she was attending graduate school at MIT's Department of Earth, Atmospheric and Planetary Science, where her advisor was David Jewitt.

Together, Luu and Jewitt began scanning out past Neptune and Pluto, beginning in 1987, and for five long years had nothing to show for it. They were, however, persistent, and had faith that new developments in digital CCD technology would allow them to finally capture glimpses of the faint Trans Neptunic objects, if any were there to be found. When Jewitt left MIT in 1988 to take up a post at Manoa's University of Hawaii, Luu followed him in due course to continue their working partnership, only leaving in 1990 to pick up a post-doctoral fellowship in Cambridge's Center for Astrophysics. She continued her work with Jewitt, however, and on 30 August 1992, on the second night of using a new, more sensitive CCD camera setup, the pair found evidence of an object, known officially as 1992 QB_1, but nicknamed 'Smiley' by Jewitt and Luu in reference to the George Smiley character of John Le Carre's spy novels. Smiley was an object about one-eighth the size of Pluto, located at 44 AUs from the Sun. (Pluto lies an average distance of 40 AUs from the Sun, though it can get as far out as 49.)

At that point, the race was on to show that Smiley was not just a loner fluke, but a member of a large family which might be the key to a number of standing mysteries in our cosmic neighbourhood. Over the next twenty years, Luu and Jewitt would find a number of such objects, with the total standing (as of this writing) at thirty-seven minor planets. The existence of the Kuiper Belt provided an explanation for short-period comets, i.e. comets with an orbital period that can be measured in a few dozen years. These objects most probably originated in the Kuiper belt, when contact with Neptune's gravitational field gave them a jolt that sent them streaming as comets towards the Sun. For their work on discovering the first inhabitants of the Kuiper Belt, Luu and Jewitt were awarded both the Shaw Prize and the Kavli Prize in 2012.

Since 1992, astronomers have found the Kuiper Belt to be a rich preserve of astronomical objects of many varieties, including the dwarf planets Sedna (discovered 2004) and Eris (discovered 2005), and the planetoid FarFarOut (discovered 2018) which currently holds the record for the farthest known object in our solar system from the Sun (at a distance of 132 AUs, or roughly three times further away than Smiley). From the relatively staid and known solar system which they had inherited, Luu and Jewitt gave us a whole new realm of possibility to explore, where astrophysicists and planetary astronomers have let their brilliance run wild, proposing new worlds from the gravitational clues that lie on beyond Neptune, with Golden Age science fiction names like Planet X and Planet Nine. In one day, our little corner of the universe got a lot more interesting, and a lot more fun, and we owe a good deal of that to an immigrant and war refugee who never stopped following wherever her curiosity might lead.

SPACE EXPLORATION

Chapter 27

The Women's Space Programme that Wasn't
The Story of the Mercury 13

The story of the Mercury 13, as it is usually told, features thirteen heroes, two villains, and a throng of supportive roles stretched between them in the grey. It is the story of a young institution repeatedly doubling down on shaky starting premises, a few highly placed politicians damning a worthy project through inaction, and one profound betrayal that thrust defeat on the shoulders of near victory.

It will make you very angry, or at least it should, as an example of a nation rewarding bravery with humiliation.

But it also offers the consolation, or at least it might, that scorn and mockery, however thickly applied, can result in the awakening of mighty forces that nudge the world to achieve its best instincts towards fairness and equality.

For three years, between 1960 and 1962, there was a hope, though slim, that NASA would put energy and resources behind a women's space programme. When the space race began on the heels of Sputnik in 1957, an arbitrary decision was made that American astronauts would come from the ranks of military test pilots. Test pilots were accustomed to speed, complex instrumentation, and the stress of working on the fringe of the physically possible, it was reasoned, and so would make the best pioneers for humanity's journey into space. The problem, of course, was that because of restrictions of access, military test pilots were all men.

America possessed dozens of women pilots with thousands of hours of flight experience across dozens of different single and multi-engine aeroplanes, but they were kept strictly away from the testing of military jets, and therefore were not among the candidates for the original Mercury programme. As a result, for two years, while NASA was running the male astronaut hopefuls through batteries of physical fitness tests, collecting invaluable data about how the male physiology responded to different stresses and strains, it was learning precisely nothing about how women might respond to those same conditions.

One arbitrary decision led to another in a chain of causation that made it more and more difficult to argue for women's inclusion in the space programme. Because only men were tested, only men's physiologies were known, and because only theirs were known, all of the equipment being developed was specific to male dimensions and needs. A devilish ratchet was in the works – the more patient women were about delaying their entry to the space programme, the further NASA went in developing male-specific technologies, and the easier it was to deny them entry into the space programme because, after all, all of the devices were made for men, and it would be a shame to go back and start all over again.

It was clear that women's patience was being fashioned into a weapon against their ever joining the space programme, and that somebody needed to step forward and begin the process of actually determining how women stacked up against the male astronauts in physical and psychological testing to establish something like a baseline for comparison. Fortunately, the first person to rigorously do so was also one of the leading names in aerospace health and medicine, Randolph Lovelace. He was NASA's go-to expert on medical testing for potential astronauts, and he believed passionately that an opportunity was being missed in not testing women. Might women not have physical properties in terms of body mass, oxygen consumption, and resistance to extremes of temperature and motion, that might make them better candidates than men for the space programme, and if so, should we not find that out sooner, while the system is still developing itself, rather than later, when we will already be too committed to a less-than-optimal system to efficiently reform it?

The official bureaucracy was not acting to correct the massive gap in its knowledge base, so Lovelace decided to start investigating the issue independently using the resources of his Lovelace Foundation, which had been intimately involved in the testing of the original Mercury astronaut candidates. Working with General Donald Flickinger, he chose one of America's most famous women aviators to begin secret preliminary testing: Jerrie Cobb.

An Oklahoma native, Cobb had spent her life clawing her way into aviation history. A craver of solitude, she undertook long cargo journeys across the globe in all manner of single- and twin-engine aircraft, racking up 10,000 hours of flight time that gave her more cockpit experience than any of the male astronauts chosen for the Mercury programme (though no jet experience), and she held three different world records for speed, altitude and distance by the time Lovelace and Flickinger tapped her as their first experimental subject. Over the next two years, it would be Cobb who led the cause of the Mercury 13 to the office of the vice president and the Halls of Congress, and after it was all over it would be she who drifted furthest afield as she attempted to put the frustrations of the past behind her.

In February of 1960, however, Cobb had nothing but an exciting future ahead of her as she began her testing at the Lovelace facilities. Tough, practical, and capable of unparalleled feats of concentration, she performed as well on the medical tests as the Mercury astronauts, and surpassed them in some respects. Her case showed that women could not only get by as astronauts but might even excel in ways that NASA had not even bothered to test. Excited by the results, Lovelace began planning to extend the testing to more candidates, to prove that Cobb was not a mere statistical fluke. He assembled a list of likely women aviators, and asked Cobb for her recommendations for more.

And he brought one other person into the project, an addition that would prove darkly fateful to the women's space effort. Jacqueline Cochran was at the time perhaps the world's most famous living woman aviator and a genuine hero. She had organised the WASPs during the Second World War to provide crucial women pilots for the war effort, was the first woman to break the sound barrier, and held more air records than we can possibly recount here. She was a superstar whose marriage to a grotesquely wealthy and influential industrialist, Floyd Odlum, gave her access to resources that most women did not have, most particularly hours in the cockpit of various military and civilian jets. Unfortunately, she did not see her privileged possession of these opportunities as a sign of something systemically wrong with women's access to the higher echelons of aviation, but instead tended to denigrate women who could not obtain those restricted resources as Not Wanting It Enough.

Her fundamentally conservative approach to governmental structures would eventually doom the Mercury 13 and set women's space flight back two decades, but at the start her influence was largely positive. She made suggestions for pilots, and offered to cover travel expenses for the candidates travelling to the Foundation for the tests. Cobb and Lovelace narrowed down a field of some 800 candidates to a couple of dozen, Cochran paid the bills, and the first round of testing began, at the end of which twelve women, ranging in ages from 21 to 41, proved themselves the equals of the original Mercury 7.

The press went mad for the women aviators who were equalling and besting the military's best and brightest, but NASA remained unmoved, citing the fact that they were too deep into designing systems for the men to consider putting aside resources to even test women further, let alone develop equipment for them. Undeterred, Cobb decided, on her own initiative, to undergo the second set of trials, psychological tests that determined whether an astronaut would mentally go to pieces or not under the pressures of space flight and surrounded by the total stillness of space. One of the most disorienting and mentally challenging of tests was the isolation chamber, in which candidates were sealed in a room without light or sound and left there. The male astronauts were placed in a simple dark room which had chairs and a table and managed to stay there a couple of hours.

Cobb, and the few women who managed to come after her, were subjected to a much more thorough test – a full isolation chamber in which they floated naked in body-temperature water in total darkness and silence. Most who entered found the experience so unsettling that they left within an hour, and those who remained had a tendency to wildly hallucinate as their brains latched on for something, anything, to call real. Cobb stayed in for six hours and showed no signs of hallucination or panic, and two of the Mercury 13 who followed improved even upon that marathon of calm in the face of total isolation, putting thereby to a definitive end the myth that women lacked the psychological capacity to deal with the stillness of the cosmic void.

What was to follow ought to have been a third and final round of endurance testing to definitively determine the relative capacities of men and women as astronaut candidates. These Phase Three tests were to be carried out in Pensacola at the Naval School of Aviation Medicine. Cobb was able to take the Phase Three tests and characteristically passed them all, but just when it was time to invite the other twelve hopefuls, the entire programme was suddenly cancelled. The navy refused to allow the use of their facilities for a project that NASA was very actively not requesting.

NASA had been vigorously insisting that it had no plans for a women's astronaut programme, and that they were far too busy getting the men into space to even consider the utility of starting to consider testing women. Meanwhile, within the small group pushing for the resumption of testing, the first strains of dissension were stirring. Cochran, who many considered harboured some resentment at not being considered for the programme herself, showed an unfortunate side in her character as she saw Cobb assume the mantle of spokesperson and chief strategist for the group. While Cobb went across the country with fellow Mercury 13 member and Congressional wife, Jane Hart, arguing that testing must be allowed to resume, Cochran wrote increasingly testy letters to her long-time friend Lovelace expressing agitation at her diminishing leadership of the group and disapproval of Cobb's tactics and message. She wrote to NASA chief James Webb outlining her own position, that women should only be considered after men had accomplished the first milestones, and only if it would not be too much of a drain on their resources to do so.

Public support was with the Thirteen, but when Cobb and Hart brought their message to Washington, those who declared themselves allies went on to use precisely none of their influence to further the Mercury 13's cause, culminating in a meeting with Vice President Lyndon Johnson, who expressed sympathy in person but wrote 'Let's Stop This Now!' across a statement drafted by one of his staff urging NASA to reconsider its stance. There would be no help from Johnson, but after much lobbying there was going to be one last chance – a Congressional hearing to investigate whether the astronaut selection process had demonstrated gender discrimination, and whether efforts should be taken to begin the process of admitting women to the astronaut programme.

The chairperson was sympathetic, and on the first day Cobb and Hart combined to deliver profound testimony about the advantages of women astronauts from a scientific standpoint, the qualifications of the women who had passed the tests so far, and the fundamental unfairness of the selection criteria and arguments used by NASA and the government to date. It was a powerful opening, but was almost immediately undercut when the next speaker stepped forward: aviation icon Jacqueline Cochran. She devoted her time to downplaying the difficulties women had in accruing hours on jets, warning of the dangers of spending money to train women who would likely just get pregnant anyway, stating that women should not be allowed in the Air Force Academy, and insisting that the women's space programme be delayed indefinitely as long as the country possessed plenty of good male candidates who had already been selected.

It was devastating testimony from perhaps the most respected woman aviator in the country, and it was amplified the next day when astronauts John Glenn and Scott Carpenter reiterated her point: with men trained and available, why spend the time and money to test women? There was to be a third day, when Cobb and Hart would have had an opportunity to rebut the misconceptions and deliberate falsehoods that had been injected into the record, but the chairperson, star-struck by the presence of Cochran and Glenn, had made up his mind that he had heard all he needed to hear, and cancelled the third day of testimony.

It was over.
The Mercury 13 would never touch space.

After the final defeat in Congress, Jerrie Cobb went to South America to lose herself as a mission pilot, flying desperately needed supplies in dangerous conditions across the continent. She had 10,000 flight hours when she was chosen by Lovelace to begin the testing programme, nearly twice as much as the most experienced of the original Mercury 7. She died on 18 March 2019, at the age of 88.

Jane Briggs Hart was an aeroplane and helicopter pilot, the oldest of the Thirteen, and the wife of a Congressman. Her testimony at the Congressional Hearing caught the attention of Betty Friedan, and the two subsequently worked closely together to form the National Organization for Women in 1966. She died in 2015.

Myrtle Cagle had 4,000 hours of pilot experience at the time of the testing programme, more than any of the Mercury 7 except John Glenn. She was a flight instructor and member of the Civil Air Patrol.

Janet and Marion Dietrich were identical twins and intercontinental air race champions. After Mercury, Janet was a federal pilot examiner and commercial aviator, the first woman in America to earn an Airline Transport Pilot License. Marion died in 1974, and Janet in 2008.

Wally Funk was the youngest of the Thirteen, and a champion of the isolation chamber. She is now 83 years old, with 19,000 flight hours under her belt and over 3,000 students to her credit, as well as a long career in aviation outreach and federal inspection.

Sarah Gorelick was the only member of the Thirteen with a degree in science, a bachelor's in mathematics with a minor in physics and chemistry. She had to give up her job in order to have the time to attend the testing sessions, and upon their abrupt cancellation she became an accountant at the IRS. She died in 2020.

Jerri Sloan was Cobb's first choice to join the programme, and had flown everything from racing planes to B-25 bombers by the time of the Mercury tests. She died in 2013.

Bernice Steadman had 8,000 hours' flight experience when she joined the tests, and was the owner of a flight school in Flint, Michigan. She also earned an Airline Transport Rating after the end of the programme, and co-founded the International Women's Air and Space Museum. She died in 2015.

Irene Leverton was a charter pilot with 9,000 hours of experience when the call came, but her employer was unwilling to give her time off to undergo the testing. She ultimately lost her job as a result of trying to become an astronaut, and after a brief period of virtual homelessness, finally found a job at Hawthorne Aviation. She died in 2017.

Jean Hixson was the second woman to break the sound barrier, flew B-25s as an engineering pilot for the WASP in the Second World War, and was a colonel in the Air Force Reserves until she retired in 1982. She died of cancer in 1984.

Rhea Woltman was a commercial and derby pilot, a seaplane pilot, and eventually a glider pilot subsequent to the end of her astronaut testing. She died in 2021 at the age of 92.

Gene Nora Stumbough was the only one of the Thirteen to actively side with Cochran's programme. She served as president of the Ninety-Nines, the women's aviation group founded by Amelia Earhart, and was a Beechcraft spokespilot.

Chapter 28

Valentina Tereshkova
The First Woman in Space

'It is I, Sea Gull!'

The words, full of meaning both personal for the speaker and cultural for the country listening below, were the first spoken by a woman in space. On 16 June 1963, Valentina Tereshkova (b. 1937) broke the bonds of Earth while sitting in the spherical confines of her *Vostok 6* capsule, and in her opening lines to the planet below she broke the designated protocol for one thin moment to breathe in the exultation of what just happened. For her, the call sign, Sea Gull, harkened back to her peasant youth when she jumped off bridges into flowing streams at the daring of her companions, her arms held out wide like a giant sea bird. To her country, the voice represented a core cultural belief that women could and ought to work alongside men in all walks of life, from the university to the factory and now, outer space itself.

After returning to Earth following forty-eight revolutions of the planet over seventy hours and 1,222,014 miles travelled, she was caught in an impulsive bear hug by a beaming Premier Nikita Khrushchev who took the opportunity to gloat that, while in the bourgeois West women were still being kept from opportunities to excel, here in the Soviet Union stood living proof of communism's egalitarian ideal. It was an effective blow to the gut of NASA administrators who had completely failed to support the training of the Mercury 13 women's astronaut programme, but as an expression of the nation's ongoing commitment to gender parity in space it was to prove lacking. Valentina Tereshkova, hero of the nation and the world, would never go into space again, and the next Soviet woman would not set foot in the final frontier for another nineteen long years.

Tereshkova's is a tale of improbabilities adding up to one great certainty, and in that respect is not all so different from many stories of the middle Soviet era. She was born in the small village of Maslennikova in 1937, the middle child in a family like any other. Her father was a tractor driver and mechanic on the village collective farm, where mechanised tools were a new thing indeed and electricity still an undreamed-of luxury. When Germany invaded Russia in 1941, her father joined the army and soon thereafter died on the frontlines, leaving his wife, Yelena Fedorovna, to raise three children and work the farm by herself as best she could.

Yelena, however, was tough as they came and managed for four years before striking out for the nearest city, Yaroslavl, in 1945. That city, even in its war-damaged state, was a revelation to young Valentina, so used to the leisurely pace of rural Russia. The sounds and smells of cars and factories stimulated her young mind and her first ambition, held in the face of universal mockery from peers and elders, was to become a train engineer when she grew up. She did some research and found out there was a railroad technology school in Leningrad and, upon graduating from high school in 1953, asked her mother permission to put in an application there.

Yelena rejected the idea outright, stating that the idea of her tiny daughter operating a massive train was ridiculous, and offering instead a future of work and night study. Deflated but determined to succeed in this next stage of her life, the 16-year-old hit the Soviet job market searching for work that could contribute to the family income while also allowing her time to further her studies at night. She enrolled in the romantically titled Young Workers' School No. 10 to take technical classes and found work as a stripper in the equally majestically named Yaroslavl Order of Lenin Tyre Factory. Her job was to take rubber-tread-coated lengths of fabric and cut them to a particular length before preparing them for vulcanisation. It was hot, hard, loud work in a stifling factory environment, and what was worse, the distance between her home, her school and her work was such that even her profound reserves of energy were running perilously low by the end of 1954.

Fortunately, her mother and sister had steady jobs at the Red Perekop Factory No. 2 and were able to bring Valentina into the fold. The commute was much shorter, the machines less loud and hot, and best of all, the entire place did not smell of burning rubber. She worked at an array of ribbon winding machines while taking classes from the secondary school, which expanded her technical knowledge and would ultimately make her an attractive candidate for the women cosmonaut programme. At night she learned about electrical engineering and metallurgy, machine design and manufacturing principles, and, in what spare time she had, she pursued recreational activities with the Young Communists' League.

The greatest of her off-time activities, however, in terms of its impact on her destiny, was her decision in 1958 to join the local skydivers' association, which was part of the Air Sports Club. After six months of training, she carried out the first of what would be 125 career jumps in May of 1959. She worked as hard at her jumping technique as she did at her technical studies and had risen to a first-class rating within a couple of years, mastering water landings, delayed parachute openings, and auxiliary parachute landings – just in time for the announcement on 12 April 1961 that Yuri Gagarin had reached space aboard the ship *Vostok 1*. Humanity had placed its first representative in space, and he was a Russian.

To understand why Tereshkova ended up where she did, we first have to talk a bit about the *Vostok* capsules and what they uniquely required of the cosmonauts inside. Unlike the American Mercury capsules, with their precisely engineered conical shape that allowed for controlled re-entry into space, the Soviet capsules were essentially giant spheres completely lacking in control surfaces. As such, unless a cosmonaut was feeling particularly courageous, standard procedure called for the capsule occupant to wait until their sphere was about 4 miles above the Earth, blow the hatch above them, and then parachute down to the Earth's surface.

Advanced skydiving skills, therefore, were a major consideration in the choosing of potential candidates, and became a key element in Tereshkova's eventual selection. After the successful launch of *Vostok 2*, she wrote a note to the Supreme Soviet volunteering herself for cosmonaut training. It was a bold, independent move that happened to coincide with the 22nd Congress of the Communist Party of the USSR's decision to include women in the space programme as both a matter of adherence to communism's core ideal of equality and, as a delightful bonus, as a means of adding another Space First to the Soviets' growing catalogue and thereby kick a bit more sand on the panicking US programme.

Soon thereafter a man arrived from the government purportedly to observe Valentina's air sports club but really to talk to her and assess her qualifications to join the core of women cosmonaut candidates the Soviet Union was gathering. Four hundred likely candidates had been selected, after which testing was carried out at the Scientific Research Institute of Aviation Medicine just outside of Moscow. Valentina went through the same set of tests as Gagarin had, and passed them all, making the cut to the final twenty-three, of which four would be selected as possibilities for an upcoming *Vostok* double launch.

On 2 March 1962, Valentina left Yaroslavl to begin her training, a rigorous programme of physical and mental preparation that would, in the space of just over a year, turn her from a promising, technically minded factory worker and amateur parachutist into an internationally acclaimed cosmic pioneer. The classes she took included astronomy, space biology, astronautical engineering, celestial mechanics and geophysics. The new challenges she had to face comprised learning to pilot an aircraft and summoning the will to withstand the various high-G and psychological torments inflicted on that first generation of cosmonauts, including days inside an isolation chamber and brain-bruising minutes being whipped around in a device known whimsically as 'the devil's merry-go-round'.

Finally, on 16 June 1963, the moment she had been training so intensely for arrived. Originally, the *Vostok 5* and *Vostok 6* launches were to feature two women cosmonauts launched within a day of each other, but in March the plans changed and a male cosmonaut was launched in *Vostok 5* followed two days later by Tereshkova in *Vostok 6*. Because nobody was entirely certain of how a woman would react to space, *Vostok 6* had two planned timelines – a one-day mission if it looked as if Tereshkova's health was in danger, and an extended three-day mission if everything was looking good. As it turned out, except for some early drowsiness, Tereshkova's response to space was within the range experienced by the earlier male cosmonauts, and the three-day mission was a go.

During her time in space, she became the first person to film the Earth's aerosol layers for analysis back on Earth, and performed a series of mental tests to show how confinement and outer space affected the ability of an individual to function in orbit for extended periods of time. In her forty-eight revolutions she logged more space time than all of the American astronauts had put together, and set the women's record for solo mission length that has yet to be surpassed. At mission's end, she re-entered the Earth's atmosphere, and at the standard 4 miles above the surface of the planet ejected from her capsule and parachuted down to the surface in Kazakhstan, to be greeted by a group of peasants who treated her to a traditional welcome meal, which was, by her account, a welcome departure from the tubes of space food she had been consuming for the last three days. At the site of her landing there currently stands a statue of Tereshkova to commemorate her feat.

Prior to *Vostok 6*, Tereshkova's training as a cosmonaut was a tightly controlled state secret. Valentina was not even permitted to let her own mother know what she was doing. But upon her return, the ex-rubber factory employee was a figure of international renown, a hero to Soviets and space enthusiasts everywhere. The next year of her life was consumed with a world tour, discussing the details of her mission and the future of women in space, and meeting ecstatic crowds wherever she went. In between the punishing public relations appearances, however, Tereshkova found time to get married to fellow cosmonaut Andrian Nikolayev, a taciturn man whom she would ultimately divorce in 1982, and to have a child, Elena, in 1964.

She had not, however, given up on space. The Soviet rocket designers were dreaming big on the heels of their early successes and the Moon was in their sights. Tereshkova took it as a matter of course that she would, some day soon, be travelling there, and while she was waiting, she began studying aeronautical engineering intensely, earning her engineering degree in 1968 and her doctorate in 1977. Unfortunately, after the *Vostok* missions, the Soviet space programme encountered a series of setbacks, each of which pushed Tereshkova further from her dreams of ascending into space again.

The American space programme was, after some initial false starts, finding its feet and devoting massive resources to catching up with the Soviets, while the death of the space mastermind Sergei Korolev in 1966 left behind a leadership vacuum that made it easier for the Soviet government to cut funds to its space programme. On top of all that, a series of high-profile cosmonaut deaths throughout the 1960s made space officials keenly anxious about risking the celebrity cosmonauts they had left. Gagarin, Seryogin, Dobrovolsky, Volkov, Pastayev – all cosmonauts killed in the line of duty within a decade, and the powers that be were determined that Tereshkova, one of the space department's most popular and successful global ambassadors, would not be added to the list.

For the next decades, Tereshkova was effectively grounded, given work as a teacher of future cosmonauts, as an ambassador and hostess for visiting American astronauts, and as head of the Committee for Soviet Women. In 1979, she passed the medical examinations for the new crop of cosmonauts, but was ultimately not selected for space flight. The second woman in space would be test pilot Svetlana Savitskaya in 1982, who went on to become the first woman to go into space twice, and the first woman to perform a spacewalk.

A victim of her own success as regards space, the second half of Tereshkova's ongoing career was filled with politics and feminist advocacy, including election to the State Duma in 2011 and 2016. At 82 years old the frantic pace she set as an unknown factory worker scrabbling for an education and enough money to pay the family's bills continues and though a John Glenn-like autumnal return to space is looking increasingly less likely, we as a species will always have that voice in our ears, looking down on the Earth turning below and declaring:

> It is I, Sea Gull! Everything is fine. I see the horizon – it's a sky blue with a dark strip. How beautiful the Earth is.

Further Reading

One of the truly frustrating things in the history of spaceflight is the neglect of English-writing historians of the Soviet space programme generally, and of Valentina Tereshkova in particular. The main source is still, nearly half a century later, *It Is I, Sea Gull: Valentina Tereshkova, First Woman in Space* (1975) by Mitchell R. Sharpe, who was also a biographer of Yuri Gagarin. In 1993, Antonella Lothian published her remembrances of Tereshkova in a volume I have never been able to find, and that is pretty much it, apart from a few children's books.

Chapter 29

Scheduling for Success, Preparing for Disaster
NASA Flight Controller Marianne Dyson

In between landing on the Moon in July of 1969 and launching the first space shuttle into orbit in April of 1981, NASA learned a few things about contingency planning. Margaret Hamilton's software had saved Apollo 11's Moon landing by crafting specialised bits of code to compensate for all the different and unusual demands that might be placed on the mission's computer systems. One year later, the explosion of one of Apollo 13's oxygen tanks created a crisis that it took every ounce of mission control's ingenuity and the astronauts' stamina and training to overcome. What was clearly needed was a set of disaster plans that did for the mission in general what Hamilton's code had done for Apollo's computers.

In order not to be caught unprepared again, NASA threw itself into creating contingency plans for every conceivable failure that could strike its newest brainchild, an entire fleet of reusable shuttles. That dedication to anticipating and planning for the worst brought forth a new breed of scientific professional who had to combine in one person the talents of an engineer, psychologist and efficiency expert; somebody with imagination enough to see what *might* go wrong, rigour enough to develop procedures to deal with those failures, and insight enough to realise how those procedures might need to be amended to account for crew behaviour. And while Sally Ride was training to become an astronaut in a swirl of media attention, a red-headed Ohio woman in her twenties with just that combination of talents quietly joined NASA to make a bit of history of her own.

Marianne J. Dyson is proof that there is more than one path to a career in space. A self-professed science fiction nerd as a child, she was particularly inspired by Robert Heinlein's *Starman Jones*, the tale of an orphan who becomes a space navigator, and by the example of Uhuru on *Star Trek*. She grew up to the sights of NASA's Golden Age, the Mercury and Apollo programmes, and the confident predictions that, by the twenty-first century, mankind would inhabit the Moon, Mars, and possibly beyond. She wanted to study astronomy with an eye towards working at NASA but the divorce of her parents, friction with her new stepmother, and the financial difficulties of her mother and stepfather made college a narrow proposition. By virtue of odd jobs, ramen noodles, and living in a run-down, vermin-infested apartment, she just managed tuition, and studied physics at the University of North Carolina.

She transferred to Rice University for graduate work but found that her mathematical background always lagged a few steps behind what was necessary for her coursework. Fourier Analysis, for example, which is today a standard part of undergraduate differential equations courses, was something she somehow managed to graduate without having studied. This led to a series of embarrassing incidents in class and in discussions with teachers when she was told, quite simply, that her level of maths was insufficient to take the courses she needed to get her advanced degree. She compensated by taking more hands-on, less theoretical, courses, and learned computer programming, while trying to figure out whether to press on in academia or try for the private sector.

It was not an easy decision. At the time, she had dreams of being an astronaut, but to do that she needed either military experience or a PhD. To leave graduate school was, then, to possibly give up all hope of becoming an astronaut, while staying presented the prospect of more grinding years of always being two steps behind all of her fellow students mathematically. 'I didn't want to spend my life in a library looking up numbers or deriving equations,' she explains in her memoirs. 'Also, I was tired of being poor, living off Ramen noodles and Kraft macaroni and cheese.'

She left school and took a job at Computer Sciences Corporation, putting to use the programming knowledge she had picked up in her physics studies, while waiting to hear back from NASA about any low-level openings. She did not have to wait long – after a week of work, NASA contacted her and on 8 January 1979, she began a career that packed a lifetime of exhilaration and exhaustion into a frantic half decade.

At the time, NASA was gearing up its plans for the new space shuttle fleet, which it hoped would be able to launch one flight a month, earning enough in private satellite launch contracts to make up for the decrease in governmental funding it had known since the end of Apollo. To maintain such a punishing mission rate required a new crop of mission specialists to prioritise shuttle tasks and compile master timelines that could adapt to the unavoidable malfunctions that came with launching the most complicated object ever fashioned by human hands into space and bringing it back again twelve times a year.

Dyson was originally brought on to aid in the development of those timelines and to programme the software that was responsible for generating them, but her job profile grew dizzyingly almost from the start. She was put in charge of determining emergency procedures for the failure of crucial cooling and positional components, which meant taking ideas from engineers and astronauts, creating programs to test their efficacy, and spending long hours in simulators to test out different solutions until she had an executable and effective set of protocols for each of the failure cases assigned her. Because of her, and the other scientists and engineers working on malfunction protocols, the astronauts who went up in the shuttle would be well provided with a set of procedures for the failure of just about any component onboard.

Meanwhile, Dyson was training for a position as Flight Activities Officer, or FAO. These were the people responsible, during the flight, for maintaining and adjusting the timeline as various emergencies arose. It was a high-stress job that required a cool head, level judgment, a willingness to listen to your support staff, and the ability to combine their advice with your own experience to make on the moment decisions about how to direct the astronauts and get the mission goals done in the face of cold, mischievous reality.

Three women were training as mission FAOs at the time: Carolynn Conley, Cheevon 'Mi-Mi' Lau, and Dyson. Conley was the first up, working as FAO on the ascent leg of the space shuttle's third flight (STS-3). Dyson worked as Timeline support on that mission, and both endured gruelling schedules that involved 13-hour shifts followed by 10-hour rests followed by more 13-hour shifts. Every malfunction required lightning recalculation of orbits and landing sites, and adjustment of the main timeline. Crew sickness, malfunctioning equipment, unnoticed errors in the schedule, all combined to make the FAO position a hectic but rewarding ride.

Dyson was FAO for the Entry team on STS-4 (27 June–4 July 1982), which launched the first secret Department of Defense payload, and Lau would be FAO Entry for STS-6, *Challenger*'s first flight. They handled everything that space, machinery and man could throw at them and were given generally favourable evaluations by their superiors, yet they noticed that, in contravention of the policies for FAO training that NASA had laid out, men who did not meet the criteria were being given FAO positions, while they were being seemingly sidelined, given fewer opportunities than their training had earned.

They brought the problem up with management, and received the reply that, to have enough trained FAOs ready for the punishing monthly launch schedule NASA was aiming for, they needed to get more people in the FAO chair sooner, instead of cycling through the same three people over and over again. It was a plausible enough answer in that era when NASA was scrambling to staff its ambitious new programme, but it still had a distinct whiff of gender discrimination, and Dyson for one felt that having brought the complaint in the first place worked against her the rest of her career.

Even with the addition of the new officers, it would not be long before Dyson would have her chance to act as lead FAO, on the upcoming STS-11 and STS-14 missions. However, she had long been hoping to have a child and, in June 1983, she gave birth to a boy, Thomas. Previously, Cheevon Lau had had a baby and returned to work promptly in six weeks. Dyson wanted more time with her newborn, and requested paid leave until October, and received it.

And that is when things started going a bit sideways. She decided that four months was not enough time, and requested an extension of her leave to November. That was also granted. Then, after discussing the matter further with her friends and family, and running through the financial ramifications with her husband, she asked for further extension until January 1984. She had already handed over her lead position on STS-11, but expected to still be on for STS-14. Management, however, said they could not do without their lead FAO until January on a project that was going into dry runs already in April. If she wanted to keep that project, she would have to come back by December. She thought of her job, thought of her child, and decided that the extra time with Thomas was worth missing out on being FAO of STS-14. She declined the December return date, and her place was filled by the next in line.

She still, however, had her regular job at NASA, which they legally had to hold for her until June. She declared that she was going to come back in April, but when that date rolled around babysitting issues arose and she did not actually return until May. She had work to do, particularly on the upcoming Spacelab 3 project, but it was not the dense and meaningful caseload she had grown used to before her son's birth, added to which was a marked coolness from some of the new women at NASA.

It was an unfortunate state of affairs all around. If only NASA had had a set policy for maternity leave in place so that everybody knew what could and could not be expected. If only Dyson had requested the full amount of her desired leave at once so that management could plan accordingly, instead of inching from October to November to January to April to May. Dyson was forced to make up the rules as she went, feeling out the boundaries between NASA's manic work schedule and the needs of being a mother. It was pioneer work, necessarily unsure and stumbling, and if in the long term it benefited NASA by making them realise that not *every* mother could or wanted to be back in six weeks, in the short term it frustrated a chronically overworked management and created resentment among the women workers who felt that their chances for being taken seriously were being negatively impacted by Dyson's inability to stick to a return date.

Once back at work, Dyson soon realised that, to do right by her child, she needed to work part-time. She put in a request in October to shift to a 24-hour work week, and was told that her designation required full-time status. She decided to leave NASA and seek out work in private industry. In five years, she had risen from a timeline programmer to one of NASA's first women Flight Activities Officers. She had stood in Mission Control and made decisions that affected the success of the world's most technically complicated scientific endeavour. And now, at just 30, that part of her life was suddenly over.

She found part-time work as a consultant at a firm that was thoroughly accommodating when her second child came, but that work was a far cry from her time at the beating heart of NASA, capped by one of the saddest scenes I have ever had to read for this series. Germany had wanted to start up a Spacelab programme of their own, designated D-1, and part of Dyson's job was to use her expertise on Spacelab 3 to help them out. When their project finally launched aboard the *Challenger* on flight STS-61A (or what sane people call STS-30), Dyson returned to NASA, but not to the Mission Control that had been her home for that half decade. Instead, she sat in the Customer Center with her German clients, her only job being to translate NASA's ceaseless strings of acronyms for the Germans so that they could decide what to do with it.

In her memoirs, Dyson attempts to put a happy face on the experience, but there is no escaping the basic sadness of a woman who had once stood proudly on the floor of Mission Control, sitting in a customer service side room, translating acronyms.

Two months later, *Challenger* flight STS-51L/25 exploded on national television, claiming the lives of Francis Scobee, Michael Smith, Judith Resnik, Ellison Onizuka, Ronald McNair, Gregory Jarvis, and America's teacher-in-space, Sharon McAuliffe. Marianne Dyson has spent the three decades since *Challenger* writing children's stories and magazine articles trying to rekindle the need to explore space which was such a fundamental part of her childhood. One of her favourite projects is a permanent telescope station on the dark side of the Moon to act as both early warning system for Earth collision events and top-notch observatory. I think we can all agree, that is a thoroughly epic idea.

Further Reading

For information about the development of the Space Shuttle programme, my favourite book is Dennis Jenkins' *Space Shuttle: The History of Developing the National Space Transportation System, The Beginning Through STS-75*. It is lavishly illustrated with diagrams of just about every component and iteration you could think of, with detailed accounts of every system and flight. Dyson's autobiography, *A Passion for Space: Adventures of a Pioneering Female NASA Flight Controller* (2016), is available through Praxis Publishing and gives a real toggle by toggle feel of what running simulations and working in Mission Control during the most hectic days of the shuttle era was all about.

Chapter 30

First
The Astrophysics and Astronautics of Sally Ride

Heroes are supposed to be mono-dimensional, startling and exceptional in one narrow aspect of life and a complex, barely functioning mess when it comes to everything else. It makes us comfortable as normal humans – 'Well, I might not have written *Der Ring des Nibelungen*, but at least I'm not a serial adulterer anti-semite who can't not wear silk.'

And then there's Sally Ride (1951–2012), the professional level tennis player, Stanford astrophysicist, T-38 jet pilot, giant robotic arm jockey, double astronaut, international educator, and successful multimillion business owner who made a nation of girls dream of space flight and defined NASA's operational goals for a generation, all while hiding a multi-decade lesbian partnership from the press and, at the end, silently battling pancreatic cancer. She's Captain Marvel, Steve Rogers and Tony Stark all rolled into one 5 foot 5 inches tall bundle of concentration and purpose whose absence at the rudder of American space flight is palpable.

How did she do it? How did she succeed at everything she undertook to touch? Her upbringing certainly helped – her family was of long Norwegian stock who supported her in everything she did without smothering her in that support. They were not a family that said 'I Love You' but they would advise her to keep going when her teachers said that science was not something she ought to be interested in, and they drove her all over the state in support of her burgeoning tennis career when it became clear that she had a preternatural gift for intelligently controlling the direction of a match.

She was a sports freak who was also a first-class nerd, and instead of making a choice between one or the other, she elected to dive into both, studying astrophysics in college while competing in the nascent world of women's tennis, marvelling both at man's landing on the Moon in 1969 and at Billie Jean King's defeat of Bobby Riggs in 1973. Borders were expanding, but the question was, what could Ride's place be in the evolving revolution?

One of the legacies of her upbringing was a tendency to emotional compartmentalisation, to solving problems through a thorough and bullet-pointed analysis of the issues at hand and a dedication to solid work, repeated until all is perfect. That combination of emotional control and capacity for perfectionist repetition made her an ideal astronaut and administrator, but it came at a cost. In her private life, relationship after relationship failed at the point of her absolute reluctance to discuss feelings and her tendency to make people subservient to the latest Mission. In academic life, as a researcher in astrophysics, her advisors saw her as absolutely thorough and conscientious, with a genius for carrying out the experimental suggestions of others, but with a limited ability to make her own imaginative leaps or to propose fresh avenues of potential research.

Her time at Stanford University included work with Dr Arthur Walker on the problem of modelling X-ray absorption by interstellar gas, and research with the newly developed free-electron laser. An FEL accelerates electrons to nearly the speed of light in a magnetic field, ultimately producing a fine-tuneable bunching of those electrons into high-energy bursts, which research would become the focus of Ride's post-astronaut scientific work. It was cutting-edge high-energy science which allowed for the intellectual teamwork that Ride loved and a combination of challenging finesse with dynamic bang that was a part of everything she did, from tennis to business. But then, in 1977, an advertisement in the Stanford paper came to her attention, announcing the ultimate in Bang.

NASA was looking for a new generation of astronauts, and was, at last, extending its pool beyond test pilots in a search for 'Mission Specialists', who would specialise in non-piloting tasks for the newly proposed space shuttle programme. Thirty-five were to be selected, and they were accepting women applicants. Thousands applied, but Ride's combination of intellectual rigour, emotional detachment and physical discipline earned her a spot on the 'Thirty-Five New Guys' squad alongside five other equally phenomenal women Astronaut Candidates (or AsCans, in one of the more unfortunate cases of NASA's propensity for abbreviation).

Those six were competing for the title of First American Woman in Space. The Russians had placed two women in space already, Valentina Tereshkova in 1963, and Svetlana Savitskaya in 1982, the second of whom also became, in 1984, the first woman to walk in space. But to the thousands of schoolchildren suddenly captured by the notion that, yes, in America you *could* be a girl and grow up to walk in space, none of that mattered. The press swarmed the women candidates while largely ignoring their male counterparts, giving them all a taste of the invasiveness whoever became the First could expect as a matter of course for the rest of their lives.

But for the moment, it was to work, memorising the thousands of switches on the shuttle, running through emergency protocols and simulations, physical training, jet piloting, specialised academic courses on engineering and orbital physics, a year's worth of intense dedication that Ride thrived in. Her trainers discovered that she was a natural jet pilot and had an even more pronounced gift for operating the gigantic robotic arm that would manipulate the shuttle's expensive payloads. She logged hundreds of simulator hours, practising every motion and sequence that the arm could conceivably call for, like Ripley working the load lifters in *Aliens*. She was organised, detail-oriented, kept calm under intense pressure, and had full command of the most complicated and important components of the shuttle mission.

She was tapped to be the first ever woman CapCom (Capsule Communicator), the link between the astronauts on the space shuttle and mission control, and performed flawlessly there as well. And while Anna Fisher (a surgeon), Judith Resnik (an electrical engineer), Shannon Lucid (a biochemist), Margaret Seddon (another surgeon), and Kathryn Sullivan (a geologist and oceanographer) would all eventually go into space (Resnik tragically aboard the *Challenger* mission which exploded), it was soon evident that Ride would be the first.

Ride would make that trip twice, her groundbreaking flight aboard STS-7 in 1983, and again in 1984, logging over 300 hours of space time and becoming America's biggest space celebrity since Neil Armstrong. When *Challenger* exploded in January of 1986, Ride was placed on the team responsible for uncovering what happened. She discovered systemic carelessness that had been the result of NASA's decision to press for a densely packed launch schedule and fast turn-around time. Precautions had been tossed aside by an organisation desperate to spark the public's fascination again. Ride, disillusioned by NASA, decided to leave the organisation, and her marriage to fellow astronaut Steve Hawley.

She had been, in fact, carrying out an affair for quite some time with Tam O'Shaugnessy, a twenty-seven-year relationship that would only be made public upon Ride's death. She had had a brief lesbian relationship in college which she had told her close friends about, but her relationship with Tam was the first time she had been able to live on close terms with somebody for over five years. Conscious of her responsibility as a hero and her obligation to NASA to avoid scandal, she felt she could never acknowledge their relationship, and only took the step of registering as domestic partners just before her death in 2012.

Free of NASA, free of living a marital lie, she returned to academia, conducting laser and nuclear detection research at Stanford and then UCSD, while at the same time serving as an advisor to multiple corporations and governmental agencies on the future of space travel and the use of technology to monitor climate change. She had, in fact, been asked by NASA to compile a report on the redefinition of NASA's long-term mission goals, a massive undertaking that became the Ride Report. She argued strenuously for a Mission to Earth, which would use the resources of NASA to discover more about the planet we are on, and confirm the damage we are doing to it.

As if conducting high-energy laser research, defining NASA's vision for decades to come, advising the government and private sector, and keeping a secret relationship from the press was not enough, Ride became a businesswoman, starting a company, Sally Ride Science, with the goal of finding new ways to inspire girls and young women with a passion for science. In collaboration with NASA, she created first EarthKam and then MoonKam, which allowed students to track the path of lunar and terrestrial satellites and request specific photos from space in an age before Google Maps made the micro-inspection of the globe routine. She organised science festivals that brought girls into contact with female scientists and astronauts, and published a whole series of books showing the important contributions that women have made to just about every conceivable branch of STEM.

She had fit the experience of seven or eight lives into her first six decades when came the cancer. Her friends noticed a yellowing of her skin, and made her go to the doctor for an exam where a massive tumour was found on her pancreas, a location where survival, even with the most vigorous treatment, stood at 2 per cent.

And she fought.

And she lost, the laser-wielding astronaut businesswoman who no longer had the energy to stand, attended faithfully by the love of her life whom she was not allowed to marry. She had not told the press of her illness, and so it was with cold shock that the world found out simultaneously of her death, and her orientation, on 23 July 2012. She was gone, and we had not had a chance to say goodbye or express in a massive flood all the things she had meant to so many of us.

Since 1983, fifty American women have flown in space.

Further Reading

Lynn Sherr's *Sally Ride: The First American Woman in Space* (2014) is pretty much the only substantial book that has the advantage of close access to all the people Sally Ride had kept hidden away during her life. It would have been nice to have more on her scientific research, and the writing itself is sort of exceptionally average, but its account of Ride's emotional semi-paralysis and occupational efficiency are engrossing and true.

Chapter 31

Preparing for an Unknown Tomorrow
Astronaut Kathryn Sullivan and the Saving of the Hubble Space Telescope

This day, let us speak of a new type of hero, one whose life story is not told as a sum of new products invented and foisted upon an already choked world, but rather a preserver, a planner, a person whose job it is to look at a device and determine how to make it *last*. It is a commonplace that the twenty-first century is the clustered haven of disposable technology – where our grandmothers kept the same rotary phone glued to the kitchen wall for four decades, we consider an electronic device to be the epitome of hardiness if it is still mostly functioning after four years. The idea of building a device to be maintainable, of anticipating future problems and building in solution options for the future user, is seemingly impractical in an age of micro-miniaturisation and unprofitable besides, and so we do not often find ourselves telling the stories of great minds in the history of engineering maintainability.

There is one place, however, that still requires people with the unique skill set of rigour, foresight and ingenuity needed to make a device that will stand the test of decades: space. The cost of putting a cutting-edge scientific vessel into orbit is so prohibitively high that maintainability and upgradability, far from being nostalgic luxuries, are dire necessities.

There is perhaps no greater example of the value of planned maintenance procedures than the Hubble Space Telescope. The team investigating its maintenance and potential upgrading began pondering the problem in 1984, led by the two astronauts who would be responsible for any extra-vehicular work required to release it from the space shuttle scheduled to launch in August 1986: Bruce McCandless and Kathryn D. Sullivan. What was supposed to be a two-year mission to test and develop new tools and procedures for working on the Hubble in the vacuum of space, however, turned into a six-year odyssey in which every system of the new instrument was evaluated and adjusted to make it a fully maintainable space apparatus.

That journey, a wearying slog through bolt design and portable platform strut placement, would prove its worth when, shortly after launch in 1990, the HST began returning blurred images from space. One of NASA's most triumphant moments, the launching of an expedition in 1993 to correct the Hubble while still in orbit, had its genesis in that slog, and among its many heroes was the woman who was sure of nothing in college so much as that she would never become a scientist.

Kathryn Sullivan was, as a child, a creature of maps and words. She thrived on the pull-out maps in *National Geographic* magazines and stashed them beneath her bed in an ever-growing pile, and loved plotting courses for the family trips. Those maps suggested far-removed lands resounding with the clang of exotic words, and by the age of 10, young Sullivan discovered that she had an ear for picking up foreign languages. Travel, languages, it all seemed to add up to some sort of job in the foreign service, and Sullivan entered college with that objective and no other. She went to her advisor to ask how she could arrange her schedule to fit in Russian amongst the battery of other courses she had planned for herself, only to be informed that she would have to take three science courses as part of her general requirements. She argued strenuously against this clearly unnecessary side-path to her life's work, but no mere individual can stand against the monolith of departmental policy, and so she signed up for courses that seemed easy enough, led by well-reputed professors: oceanography and geology.

Thereupon a new world opened. Sullivan discovered that she could experience travel and utilise her flair for foreign language, while at the same time doing interesting new research in far-flung environments that made the workaday bureaucracy of international diplomacy a meagre shadow by comparison. Most tellingly for her future career as an astronaut, one of the parts about professional oceanography that captivated her the most was the preparation for a mission. The elaborate marshalling of resources and schedules to meet specific objectives within set material limitations struck her as a stimulating challenge where most see it as a bottomless chore. Getting things right ahead of time, preparing for contingencies, working within resource constraints – these were the factors that made missions succeed, and in cultivating them, she paved her way to the high altar of procedural rigour: NASA.

In December of 1976, in the midst of completing her PhD work, Sullivan learned from her brother that NASA was assembling a new team, the first in nine years, to service a new fleet of spacecraft, reusable vessels called space shuttles, and were actively recruiting women for the squad. As a multilingual 26-year-old scientist in good physical condition with experience in the minutiae of mission planning, she seemed a natural candidate, but after dropping her application in the post, she thought no more of it. Almost 9,000 people applied for the few coveted spots, and it was no use getting her hopes up with her unfinished PhD staring her in the face, demanding her full attention.

As it happened, however, of the 8,700 applicants, she was one of the 200 brought in during the summer of 1977 for further testing, and one of only thirty-five who made the final cut, joining the team of the self-styled Thirty-Five New Guys in February of 1978. Six women were among those thirty-five, including laser astronomer Sally Ride. Ride would go on to become the first American woman in space (aboard STS-7 in 1983), but Sullivan would earn her own place in the history books on 11 October 1984.

One of the selling points of the space shuttle was that it could act as an orbital repair yard and refuelling station for wayward or otherwise misbehaving satellites. Instead of writing off broken satellites as losses, the shuttle would allow teams to collect them and either repair and refuel them on the spot, or recapture them and bring them down to Earth. On October 11, Kathryn Sullivan and David Leestma stepped outside the shuttle and successfully demonstrated the feasibility of refuelling a satellite in orbit during a 3.5-hour extra-vehicular activity (or EVA). With that one step outside her craft, Sullivan became the first American woman ever to perform a spacewalk, and the second woman overall (Svetlana Savitskaya had performed a 3.5-hour spacewalk for the Soviet Union just three months prior).

It was a milestone that brought some gratifying attention, but not the crushing press of scrutiny that Sally Ride experienced after her mission in STS-7, and that suited Sullivan just fine. She was ready to go back into space at the next possible opportunity, and in the closing moments of 1984 she was informed that she would be on the mission that launched NASA's cutting-edge space telescope (brought to life over hard-fought decades by NASA chief astronomer Nancy Roman), the Hubble. And not only would she be part of the mission, but one of two people tasked with developing the protocols and tools responsible for dealing with any hiccups in the satellite's launch, and for allowing long-term maintenance down the line.

For most people, the question of, 'Do you want to spend the next one to two years intensely investigating the shape of universal bolt head to be used in this piece of equipment?' would elicit a tentative, 'Hmmm, let me think,' followed by a spirited leap through the nearest window. For Kathryn Sullivan, the job of evaluating the maintainability of the Hubble

Space Telescope and developing the tools and upgrades needed to make it a fully repairable spacecraft represented the intersection of her many talents: a profound attention to detail; an ability to devote intense concentration to minutiae over long periods of time; a gift for imagining mechanically feasible solutions; and the capacity to work as one of a team in spite of her vaunted astronaut status. She and McCandless went through the Hubble, bolt by bolt and system by system, and developed, with their support teams, uniform standards for fasteners to decrease the number of tools needed to service the telescope; new removable and adjustable platforms that could be operated by the feet, allowing astronauts to place themselves in the correct orientations for dealing with various maintenance issues; and new tools for handling electrical connections without damaging the wires or scraping off small flakes of metal that could damage the telescope's electronics.

They had accomplished much on the eve of the mission's scheduled 1986 launch, and many of their tools (particularly the adjustable foot platform) would become NASA standards in the years to come, but many systems were too centrally located to upgrade for maintainability, and so had to be left 'as is' in preparation for the imminent mission. The *Challenger* disaster of 28 January 1986, however, put all future shuttle missions on indefinite hold. NASA was stunned at the loss, and Sullivan put her grief into working with the widows of the *Challenger* crew to create an educational initiative in the memory of their lost family members.

There was no true way forward, however, except by pressing on with the work at hand. The launch delay for the Hubble meant that it could be thoroughly upgraded so that virtually all of its components could be exchanged and repaired in orbit. Space shuttles resumed their missions in 1988, but the Hubble was a comparatively low priority next to more lucrative commercial and military contracts, and so it was not until 1990 that STS-31 took off with the HST in its cargo hold. Once in orbit, the deployment of the Hubble began. The first solar array unfurled flawlessly, but the second appeared to be stuck and unable to fully extend. Sullivan and McCandless dashed for their spacesuits and began the four-hour wait in the airlock to wash the nitrogen out of their blood in case they were needed to exit the shuttle and attend to the malfunctioning array. Time was at a premium, between the unusual amount of fuel exhausted to get the Hubble to a higher orbit and the need for the solar panels to open to power the telescope's electronics before the batteries died out.

Huddled in a white room with no view to the outside, hindered by a biochemical process that could not be rushed, Sullivan ran through every step she needed to take to fix the telescope. She had been over every system on that instrument over the course of the past six years, knew where she would need to place her footholds, and what tools she would need, as soon as her blood cleared itself of treacherous nitrogen.

Then, suddenly, all was well.

The team on the ground came up with a software fix they beamed up to the HST, the solar array unfurled, and the Hubble was deployed while Sullivan remained in her white room, unable to watch the launching of the satellite she had come to think of in many ways as her own.

Hubble had launched, and on 29 April 1990, Sullivan returned to Earth. With the first images beamed back from the HST, however, NASA saw that something was deadly wrong, a microscopic fault in the Hubble's mirrors was resulting in hazy images. Ridiculed mercilessly on late-night television for a production fault less than a fiftieth the thickness of a human hair, NASA needed to repair the Hubble to save face. The last half decade had seen the loss of the *Challenger*, it could not see the complete failure of the HST as well. Thanks to Sullivan, McCandless, and their maintenance procedures team, however, the crew that flew to the Hubble in 1993 would find everything in place, with easy access to all critical systems, to perform a flawless repair that would see the Hubble serve for the next two and a half decades as the globe's pre-eminent astronomical instrument.

Sullivan, however, was not part of that repair mission, and soon thereafter she began splitting her time between her responsibilities as an astronaut (she would go to space again in 1992 aboard STS-45), and a new position at the Navy Reserve which allowed her to reconnect with the oceanographic work that was calling to her anew after a decade of being consumed by the problems of space and space exploration. In May of 1993, she was appointed Chief Scientist of the National Oceanic and Atmospheric Administration (NOAA), and retired from her work at NASA. In 2013, she was appointed by President Obama to serve as NOAA's Administrator and also as an Under-Secretary of Commerce for Oceans and Atmosphere after a decade spent developing initiatives to improve science education in the United States.

While Sullivan settled down to life on Earth and the massive responsibilities of measuring and protecting its limited resources, the Hubble went on to a life of its own. Like Theseus' ship, over the years it had nearly every component traded out and upgraded at one point or another, new experiments swapped in while old ones were discarded, until all that remained of the original telescope was a casing and some mirrors. Very little of that would have been possible with the instrument Sullivan first beheld in 1984, with its inaccessible connections, unlabelled components and tucked away bolts. In the face of engineers who insisted that advances in maintainability could not be made, or were not important enough to be made, Sullivan and McCandless relentlessly pursued a better way forward, and produced an instrument that flawlessly served the needs of a full generation of astronomers. Sullivan would have been a hero if she had never laid eyes on the Hubble, but the concentrated work she did over half a decade to do the job she had been given as thoroughly as it could be done elevates her to something beyond that title. She was no mere hero – she was that greater, rarer thing – a damn fine engineer.

Further Reading

Kathryn Sullivan released her memoir, *Handprints on Hubble*, in 2020. As an inside view into the technical aspects of mission preparation, and in particular in the long history of evolving the Hubble from a telescope that would be brought down to Earth for servicing from time to time into a fully in-orbit-repairable vessel, it is without parallel.

Chapter 32

Microgravity, I Say Thee Nay!
Space Biomedical Engineer Mamta Patel Nagaraja, The Woman Behind Women@NASA

When it comes finally time to settle Mars, the most terrifying sound might not be *RRRIP* or 'I'm sorry, Dave, I don't WANT to turn the carbon dioxide scrubbers back on.' It might well be *SNAP*. That is, *if* we cannot figure out solutions to some of the issues astronauts face in long-term space flight.

It will take astronauts nine months to get to Mars using current technology, every month bringing with it a sure and steady loss of bone mass, as osteoblasts, the cells that are supposed to lay down new bone deposits, slowly stop showing up for work. As a result, after touchdown, astronaut bones could break more easily, with the nearest hospital 249 million miles away. If humanity is to become a star-dwelling race, we have to first come to grips with the quiet treachery of our skeleton.

If you are like me, you probably thought that bone loss was a thoroughly Fixed Problem. Just give the astronauts some calcium supplements, some spring-based exercise equipment, maybe one of those spinning space stations you get on Arthur C. Clarke covers – problem solved.

Well— not quite.

To find out why, we need to look at the work NASA scientists have been doing on the physiological effects of microgravity, which happens to be the speciality of biomedical engineer Dr Mamta Patel Nagaraja. You might know her as the head of the Women@NASA programme, a magnificent resource for demonstrating the breadth and scope of the women scientists and engineers and communications experts who work at NASA, and we will certainly talk about that in a bit, but before she was the woman behind Women@NASA, she was a triple crown engineer pushing forward our knowledge about the deep mystery of how human organisms fare in space.

She was born in California in 1979 to a pair of hard-working and deeply self-sacrificing parents who had just emigrated to America from India the year before. She was the third of five siblings, three of whom ultimately became scientists. Nagaraja's tales of her youth are straight out of a Norman Rockwell painting – the siblings toting encyclopaedias into the forest on expeditions to identify the flora and fauna they found there, or borrowing gigantic puzzles from the library to do them at home as a family over the course of weeks. The three middle children were science nerds who helped and inspired each other, and Nagaraja's ultimate career is a fantastic blending of her own early love of space and engineering with her sisters' interests in genetics and physiology.

When her parents found work running a small motel, the children were drafted in to help with the maintenance work, regardless of gender, and so Nagaraja picked up skills with jackhammers and air-conditioning units, sewer lines and refrigerators, that would serve her well when she went to college and did more or less everything you are told not to do by conventional wisdom.

The Standard Advice says find a field early in college, specialise deeply in it, and then ride that speciality through your career. But Nagaraja's interests, understandably, were too broad to be whittled down to just one field. So, she did her bachelor's degree in aerospace engineering, then switched fields for her master's to mechanical engineering, then switched *again* for her doctorate to biomedical engineering. So, for all the students out there who feel that they have to decide early, and stay with that decision no matter what happens, I advise you to print out a photo of Dr Nagaraja and write in bold letters under it, 'Change is Okay'.

Of course, it helps if you are a polymath genius who comes from a supportive family of science enthusiasts and who grew up with daily practical engineering experience as a matter of course. ...

Anyhow, getting back to those treacherous permanent residents, our bones. Nagaraja was in a position to concentrate her various skills on a multidisciplinary problem that involved merging her old love of space and space travel with her skills as a mechanical and biomedical engineer. It had been known since Millie Hughes-Fulford's work in the mid-1990s that the slowing down of new bone growth had something to do with gravity's impact on gene expression. Remember, it is not enough to just *have* the genes that allow for new bone growth. They also need to be *expressed*, which is to say the instructions have to be copied down into mRNA and then put into action. Somehow, without gravity, the expression of the genes that lead to osteoblasts laying down new bone growth was being affected, meaning that the problem was more fundamental than we had thought.

The issue with studying the effects of microgravity on humans is, of course, that there are not many humans who live in microgravity. Statistically speaking, the number of people who have lived any amount of time in space is miniscule, and the number who are currently living there positively *dreams* of minisculity. So, other ways have to be found to study how gravity influences gene expression. Nagaraja uses studies of cells held in simulated microgravity (a magnetic field can do the trick) to see how prolonged exposure to different levels of gravity impacts gene expression, and how different techniques might improve that expression.

So far, it looks like the answer might lie in including a vibrational element to the astronauts' exercise regimen in space, since cells appear to respond to vibration similarly to how they respond to the force of gravity at the level of gene expression. As a graduate student, Nagaraja worked on the benefits of whole-body vibrational platforms with Dr Clinton Rubin, and the results have been promising enough to merit consideration as a payload in a future NASA space mission.

Meanwhile, because you are doubtlessly not feeling *quite* under-accomplished enough, there are all of the other things Nagaraja has achieved in her life. She was an instructor and flight controller at Johnson Space Center, training astronauts in the operation of their shuttle. And then there is the important job of running the Women@NASA programme, bringing to the world the stories of NASA's great and long history of women researchers and workers in a way that is both exciting and inspiring to a new generation sitting at their desks and dreaming of stars. For any girl in a classroom feeling alone in her love of science, there are columns and columns of role models there to search through and identify with, a great digital edifice declaring You Belong. You can also find information about NASA's middle school mentoring programmes, NASA GIRLS and NASA BOYS.

Asked about her plans for the future (she is a mere 42 now), Nagaraja looks forward to many more years of work at NASA (including a thin chance to actually go to space herself – fingers crossed!) and then perhaps a change of pace to developing knee replacements for those who cannot currently afford one or developing inner city clean water supplies both in America and in villages in India where her parents grew up. She has trained our astronauts, helped remove a thorn from the paw of our future space colonisation hopes, and desires nothing more than to cap it all with work to improve access to clean water, in a life lived with a deep sense of commitment to employing her brain for the good of others. If you need any more proof of how much we stand to lose if we close ourselves off to the world, its hopeful and hard-working people, and their children, look no further.

Chapter 33

The Concerns of the Earth and Above
Mae Jemison's Life in Medicine and Space Travel

There is a bit of political wisdom we have lived with for half a century now, which says that spending money on space travel, while social problems still exist upon the Earth, is wasteful at best and downright immoral at worst. Get all of the people fed first, critics of spaceflight often say, then you can go into space all you want, as if the problems involved in doing the former are easily solvable by the resources tied up in carrying out the latter. The problem with that line of thought is, of course, that it assumes a nation cannot be several things at once – that advancing any one area of human concern must come at the expense of another, and that people who devote their lives to space flight must do so myopically, without appreciation or concern for humanity's other, and nearer to home, projects.

But nations and individuals alike carry multitudes within them, and there are few examples of humanistic and exploratory instincts not only residing in the same person but augmenting and elevating each other, than the first African American woman in space, astronaut Mae Jemison. By turns a medical doctor, engineer, Third World nation healthcare worker, space explorer, political activist, media personality, writer and dancer, Jemison shows that the drive to know more about our universe does not need to come from a cold place of calculating conquest; it can be the ultimate expression of artistic, intellectual and humanitarian drives all pulling together towards a more equitable future.

Mae Jemison was born in Alabama in 1956, the youngest of three siblings, but by the age of 3 her family had moved to Chicago, part of the Second Great Migration that saw 5 million African Americans move from the South to the North and West in the decades following the Second World War, seeking better job and educational opportunities. As a child, Jemison possessed a voracious desire to not only know about the world, but to *do* things in it. She devoured books on astronomy and planetary development during her weekly raids on the local library, built careful models of the Earth's evolution over time, and begged her parents for dance lessons.

A precocious and fearless youth, Jemison seemed blissfully unaware of what is and is not done in the pursuit of knowledge. During high school, when she decided she wanted to know more about sickle-cell anaemia, she simply called up Cook County Hospital and asked to be connected with the head technician of the haematology laboratory. That act of daring got her an invitation to visit the hospital, which in turn became an informal internship complete with an original research project.

At 16 years of age, Jemison graduated from high school and was accepted into her dream college, Stanford University. At the time of her entry, Stanford had only recently begun its experiment with co-ed dorms, a situation that the young Jemison took remarkably in her stride as she dived into the complicated world of higher education. She embraced the opportunity to study African history and language, engage in student activism, and play pick-up American football games with her classmates, even as she found her way often barred by unsympathetic professors who could not quite believe that a young black woman could be seriously considered as on par with the predominantly white male students in the engineering department. In spite of her chock-full schedule and the presence of institutional roadblocks, Jemison graduated from Stanford in 1977 with a double major in Chemical Engineering and African and African American Studies.

It was at this point that Mae Jemison the future astronaut very nearly became Mae Jemison the professional dancer. In the last quarter of her time at Stanford, she staged and choreographed *Out of the Shadows*, a dance production to be held at Stanford's Little Theater. The show was a success, and Jemison considered taking the next year off school to see if she could make it as a dancer/choreographer before proceeding on to medical school, a plan she ultimately dropped on her mother's noting, 'You can always dance if you are a doctor. But you can't easily doctor if you are a dancer.'

So, Cornell Medical School it was, and yet another change in perspective for the intellectually flexible Jemison. As a chemical engineering student, she was used to understanding fundamental principles and building upwards from those; if a particular value or constant was needed, it could always be looked up, and the memorisation of columns of standardised values was not considered to be a fruitful use of brainpower or time. She started as a medical student assuming the same approach would apply – know the underlying chemical principles and you should be able to derive the resulting bodily behaviours. That was, to put it mildly, not how medical school worked or works, and her first class in Gross Anatomy brought her to the realisation that she would have fundamentally to rethink her approach to this new field of knowledge, regearing herself as a memorisation machine.

That is not an easy process to undergo – the mathematical brain and medical brain make very different demands – but Jemison managed it, and further managed to combine her scientific interests with her humanitarian and global concerns, spending the summer after her second year as a medical student in Kenya, working with the Flying Doctors (African Medical Education and Research Foundation), whose goal was to bring medical services to underserved regions of East Africa. It was a formative experience that would lead to her accepting, after a year interning at the Los Angeles Medical Center, a position as the Area Peace Corps Medical Officer for Sierra Leone and Liberia. Jemison served as the director of health services for all the members of the Peace Corps in those countries from 1983 to 1985, in which capacity she not only saw patients, as a primary care doctor, but managed the facilities that made the most out of the limited resources at hand in combating a wide array of deadly tropical illnesses.

Returning to the United States, Jemison worked as a general practitioner at a Cigna facility in Los Angeles while continuing her education in engineering and business, at the same time also beginning the process of applying to the NASA astronaut programme. Just two years previously, Sally Ride had become the first American woman in space, inspiring a generation of like-minded overachievers to follow in her footsteps. Jemison sent her first application into NASA in 1985, but astronaut selection was delayed by the 1986 Challenger disaster, and so it was not until 1987 that Jemison had a chance to re-apply for the programme. Two thousand qualified applications were received in that round, of which fifteen would ultimately be selected for Astronaut Group 12, Jemison among them.

She was the first black woman to be selected as a NASA astronaut, and in 1992, she became the first black woman in space aboard STS-47, where she served as one of four mission specialists, responsible for overseeing the forty-plus experiments loaded into Spacelab J. The mission lasted 190 hours, and was Jemison's only venture into space. She retired from NASA in 1993 (the same year that she appeared on an episode of *Star Trek: The Next Generation*, marking the first time someone who had actually been in space appeared on the show). Subsequently, she devoted herself to issues involving the intersection of technology and culture. She formed Jemison Group Inc in 1993 as a consulting firm specialising in issues of how to make tech designs that are accessible to developing sectors of the world, and in 2012 took over lead of the 100 Year Starship project, which studies what resources and institutions will be necessary to allow the achievement of human interstellar travel within the next century.

Jemison continues her work on multiple fronts, as we could only expect from a person of such multifaceted talents and interests. She has coupled the fame she gained as the world's first black woman astronaut with the decades of experience she has built up as an engineer, doctor, explorer and activist to create new educational initiatives and public programmes that will bring the benefits of scientific study and exploration to a broader-based world community, while also fostering the intellectual curiosity and opportunities for academically underdeveloped parts of the world. A wider array of perspectives entering into science, working in a feedback loop with a science that looks beyond the capacities and concerns of the first world – it is an exciting prospect, and one perfectly tailored to a person like Mae Jemison to shepherd through the next decades of our grand human experiment.

Further Reading

Mae Jemison wrote a young adult memoir in 2001, *Find Where the Wind Goes*, which is an incredibly engaging account of her life, focusing on her years as a child and student. I have the 2020 2nd edition of the book, which unaccountably still contains dozens of typographical errors that Signal Hill Road Publishing deserves a hearty wag of the finger for allowing into their final product. Jemison deserves decidedly better quality control than that, as do the young people reading about her life. Still, it is a great story, and Jemison's style is eminently readable. Jemison also has about seven pages devoted to her life in Karen Bush Gibson's wonderful *Women in Space: 23 Stories of First Flights, Scientific Missions, and Gravity-Breaking Adventures* (2014).

Chapter 34

Brief Portraits
The Twentieth-Century Turn

Fiammetta Wilson (1864–1920) and Alice Grace Cook (1877–1958)

Alice Cook and Fiammetta Wilson's names will perhaps always be linked in the annals of astronomy. Both were keen astronomical observers who specialised in meteors, and with the advent of the First World War, and the resultant draining of male astronomers from the Meteor Section of the British Astronomical Association, the leadership and survival of that institution was given over to Wilson and Cook to see through the war years. From 1916 to 1918, the pair conducted their observations and saw to the regular quarterly publication of the *Report of the Observing Sections: Meteor Section* which they also wrote together.

Wilson was Cook's senior by some thirteen years, so when the story of their joint war work is told, it is often Wilson's story with occasional glimpses of Cook, but in truth their lives ran in substantially parallel directions thanks to Wilson's delayed start in astronomy. Born Helen Frances Worthington, Fiammetta's parents had the resources to provide her with a string of governesses, admission to quality schools in Germany and Switzerland, and an extensive musical training in Italy. In 1889, she married a music teacher, and within a decade the pair were separated, possibly due to her husband's philandering tendencies. In 1901, Helen changed her name to Fiammetta Waldahoff for reasons that are not entirely clear, and for the next decade she supported herself by teaching mandolin and conducting at the Guildhall School for Music. In 1907, she applied for divorce, and the following year married Sydney Arthur Wilson, who was some eleven years her junior.

Finally, in 1910, as a result of attending a lecture series by the astronomer Richard Fowler, Wilson decided to devote herself to astronomy, and joined the British Astronomical Association the same year. This sudden turn towards astronomy was almost exactly mirrored in the story of Alice Cook, who attended a lecture series by Joseph Hardcastle in 1909, and joined the British Astronomical Association in 1911.

They might not have known it yet, but the pair of astronomers were now marching in a lock step formation with each other that would not truly end until Wilson's death in 1920. They both developed an interest in meteor observation, zodiacal light (the triangular section of diffuse light visible just after sunset and just before sunrise), planetary transits, lunar occultations (when the Moon passes in front of a celestial object), auroras, and comets. From 1910 to 1920 Wilson discovered some 10,000 comets (sometimes in the midst of zeppelin raids) and calculated the paths of over 600 of them, while in 1913 Cook wrote a study for the *Journal of the British Astronomical Association* on the 1876 meteor display, and in 1915 wrote a report on meteor magnitude for *The Observatory*.

The year 1916 was a watershed one for both Wilson and Cook, as it was the year they entered history both as part of the inaugural group of five women inducted into the Royal Astronomical Society (the other three being Mary Blagg, Irene Warner and Ella Church) and as leaders of the British Astronomical Association's Meteor Section, a position they would hold until 1919. Wilson's death in 1920 is something of a mystery, one which deepens with the fact that her husband remarried only seven weeks after her death. Now, I am not saying he murdered her, but I am also not saying he did not. ... In any case, with Wilson gone, Cook continued to direct the Meteor Section by herself until 1923. Information about her life after leaving the Meteor Section is scarce on the ground, until she pops up in 1950 as a co-founder of the Ipswich and District Astronomical Society. In spite of being one of the first women inducted into that organisation, the Royal Astronomical Society did not bother to write the usual obituary on her life and accomplishments upon her passing in 1958, while the British Astronomical Association contented itself with a paragraph's worth of mention at its 25 June 1958 meeting for the woman who had kept its meteor research afloat during wartime.

Antonia Caetana de Paiva Pereira Maury (1866–1952)

I suspect that every student of astronomy has, if you press them hard enough, a favourite among the Pickering-Era explosion of Harvard women astronomers. Those prone to recognising the importance of administration might gravitate towards Williamina Fleming. The theory junkies will probably pick Henrietta Swan Leavitt, while the catalogue jockeys might lean towards Annie Jump Cannon. I suspect, however, that for the Cool Kids in the room, the choice has to be Antonia Maury. Among the Harvard group, Maury stands out as the rebel, the dreamer, the tragic hero, who was cut down because her vision was too large for the task assigned to her, and she was too confident of her worth to allow herself to be reduced to something less than she was.

Maury came from scientific royalty. Her great-grandfather was the court physician to John VI of Portugal. Her great-great-grandfather was Thomas Jefferson's teacher. Her grandfather was John William Draper, who in 1840 made the first daguerreotype of the Moon, and her uncle was Henry Draper, who was a pioneer in the study of stellar spectra, while her aunt was the driving force behind funding the comprehensive spectral mapping of the night sky that formed so much of the life's work of Pickering, Fleming, Cannon, and eventually Maury herself. Suffice to say, Maury did not lack for role models, and when she entered Vassar in 1883 it was only natural that she would gravitate towards its head of astronomy studies, arguably the most famous woman astronomer in the world at that point, Maria Mitchell.

Maury graduated in 1887, two years before the death of Maria Mitchell brought to a close the first great era in women's astronomy in the United States. The previous year, Maury's aunt, Anna Draper, had brought the Harvard Observatory the funding it needed to carry on Harry Draper's work, and it was only natural that Maury would join that effort though, for all of her distinguished astronomical lineage, she was still only paid at the 25 cent an hour rate of the other women computers at Harvard.

Pickering and Maury did not get along. In the process of analysing stellar spectra, Maury was frustrated by the broadness of the classification scheme worked out by Fleming, and devised her own finer classifications to attempt to more accurately categorise the stars she was analysing. This included the creation of a system to account for the distinctness of the observed lines, the *abc* classification, which included the important observation that some spectra feature narrow, sharp hydrogen and helium lines, and particularly sharp calcium lines. She denoted this type of spectral contour type *c*, and later, in 1905, type *c* spectra were recognised as characteristic of red giant stars. It was an important distinction that captured one of the night sky's most important residents, but during Maury's time, all of her attempts to create finer and more distinct categories just represented an irritation to Pickering, who saw her approach as gumming up the works, increasing the time it took to analyse each star during a project when speed was of the essence if the Draper Catalogue was going to be completed in anybody's lifetime.

Further, Maury and Pickering butted heads over credit for the discovery and description of binary star systems. Pickering had discovered the first spectroscopic binary star (a binary star is a system of two stars in gravitational orbit around each other, and a spectroscopic binary is such a system that you cannot see visually, but can distinguish on the basis of changes in Doppler Shift of the system's spectra) in 1887, and had handed off the complicated task of computing its orbit and period to Maury, who not only pulled off that task but discovered a second binary system herself, and calculated its orbit as well. In his report on the binary discovery, Pickering mentioned Maury's work once in relatively general terms, which Maury considered inadequate recognition of the work she had done.

In 1891, Maury had had enough of her treatment at Harvard, and left the Observatory, returning only briefly in 1893 and 1895 to complete her work – on the condition that she could use her own classification system and that she would be given full credit for the work she did. The results of that work were published in 1897 with Maury's name in the title as *Spectra of Bright Stars Photographed With the 11-Inch Draper Telescope as Part of the Henry Draper Memorial and Discussed by Antonia C. Maury Under the Direction of Edward Charles Pickering*.

With Pickering's death in 1919, and the assumption by Harlow Shapley in 1921 of the direction of the Harvard Observatory, Maury felt that she could profitably return to the Harvard Observatory, remaining there until 1948. She received a vindication for her classification distinctions in 1922 when some of its elements were incorporated into the International Astronomical Union's classification system, the broad outlines of which remain in place today.

Mary Acworth Evershed (née Orr) (1867–1949)

The last time you read the work of Dante Alighieri, the first thing that came to your mind likely was not, 'Huh, I wonder what the cosmological system being outlined here is, and what astronomical science it is based in.' That is because you are not Mary Evershed, a good chunk of whose career was devoted to determining Dante's astronomical perspective and publishing her results in her 1914 book, *Dante and the Early Astronomers*. For most, that is the extent of what they remember about Mary Evershed, but her career as an astronomer was far too vast to be summarised with the epithet 'That Dante Astrohistorian'.

Evershed had no formal schooling, being taught entirely at home by a string of tutors, and allowed to follow her many intellectual curiosities through reading and travel. From 1888 to 1890 her family lived in Florence, where Evershed was struck with the astronomical errors in Dante and decided to look further into the topic of early astronomical practices for herself. In 1890, the family moved to Australia for five years, where Evershed was frustrated with the lack of good astronomical guides to the Southern sky, and so decided to write one herself, *Easy Guide to the Southern Stars* (1896), and a more full account published after her return to Britain, *Southern Stars: A Guide to the Constellations Visible in the Southern Hemisphere* (1896).

Back in Britain, she joined the British Astronomical Association, which had been founded in 1890 by William Maunder and was open to women from the start (for more about which, see the portrait of Annie Maunder below). Here, she met some of Britain's famous practising astronomers, like Agnes Clerke and Annie Maunder, and began her own astronomical researches, which she published in the *Journal and Monthly Notices of the British Astronomical Association*. In 1906 she married John Evershed, an industrial chemist with a passion for astronomy who that year was appointed assistant director of the Kodaikanal Observatory in India. And so, it was onwards once again to new environs for Mary, as she and John stayed in India for the next seventeen years conducting solar research together. Her 1913 paper *Some Types of Prominences Associated with Sun-Spots* was, in particular, a prescient document that advocated for the importance of magnetic forces in determining the course of solar events. It was also while in India that Evershed wrote her book on Dante's astronomy, as well as an updated guide to the Southern sky, *Stars of the Southern Skies* (1915).

Returning to Britain in 1923, Mary was made director of the historical efforts of the BAA, and was inducted into the Royal Astronomical Society at last in 1924, where she similarly had a role serving on its library committee. During this period, she was primarily concerned with preserving and investigating astronomy's rich past, one of the more public results of which was her 1935 article *Arab Astronomy*. But Evershed would not be Evershed if she kept too close to the standard script, so of course in 1938 she published a work in a new field entirely, *Who's Who In the Moon*, a 130-page summary of lunar craters. It was her last published work.

Annie Scott Dill Maunder (née Russell) (1868–1947)

Annie Maunder was one of the last woman astronomers to have to fight her way through the full might of the British restrictions on women's scientific education in the nineteenth century. Born in Strabane, Ireland to a Presbyterian minister and his second wife, she was surrounded by people of learning and talent (her sister studied medicine with Elizabeth Garrett Anderson, whom we met in the previous volume of this series, and she had two brothers who took up careers in astronomy), but her early education at home was not precisely exhaustive, and upon her arrival at Girton College in 1886 she found, according to her lecturer from that time, that she had to work three times as hard as her classmates in order to overcome the gaps in her early education. Nonetheless, work three times as hard she did, and in 1889 graduated the top mathematician of her class, with what would have been a second-class performance on the Tripos exam if women students of that era had been allowed to compete for Tripos honours. Her gender also prevented her from receiving a bachelor's degree, though she had done all the coursework to have earned it.

Following her graduation, she took up work as a mathematics teacher, but upon hearing of an opening at Greenwich Observatory in the human computer pool of the Astronomer Royal, William Christie, she applied, was rejected, applied again, and was accepted, joining the observatory in 1891. It was a decided step down in terms of compensation – her salary as a teacher was £80 per year with board provided. As a computer, she earned a threadbare £48 per year, without any provision for boarding.

While at Greenwich, she met, fell in love with, and married the head of the Spectroscopic and Photographic Department, William Maunder. He had helped found the British Astronomical Association in 1890, a group of astronomical enthusiasts that counted among its starting council four women, at a time when the Royal Astronomical Society was still twenty-six years away from admitting its first woman non-honorary member. Walter was a widower with five children who was devoted to women's role in professional astronomy, but their marriage in 1895 meant that Annie had to leave her position at the Observatory after a mere four years there. She continued working in astronomy in an unofficial capacity, including accompanying William on five separate eclipse observation missions from 1896 to 1905, and writing the popular astronomy book, *The Heavens and Their Story* (1908). She also served as editor of the *BAA Journal* from 1894 to 1896 and later from 1917 to 1930.

In the early 1900s, she undertook the work which firmly established the Sun's eleven-year sunspot cycle proposed by Samuel Schwabe in the 1840s, and captured elegantly in the Maunder 'Butterfly Diagrams' of solar activity, as well as the existence of east-west asymmetry of sunspots. By 1916, her reputation in astronomical circles was such that she was included in the second round of women inducted into the Royal Astronomical Society (she had been nominated in 1892 but was rejected in a secret vote). Later in her career, she turned towards an interest in ancient astronomy, particularly in the subject of the forty-eight original Ptolemaic constellations.

In 2018, the Greenwich Observatory installed the Annie Maunder Astrographic Telescope in honour of the woman who began her career there as a human calculator, and ended it as an internationally recognised expert on the history of our Sun.

Caroline Ellen Furness (1869–1937)

'When I develop into a second Mary Somerville, or Maria Mitchell, then you will be proud of me.'

These words were written by Caroline Furness to her sister during her sophomore year at Vassar, explaining her definitive turn towards mathematics and science as her field of study. Her interest in the great women of nineteenth-century science was understandable, for at Vassar she had as her very-present role model the figure of living legend Mary Watson Whitney. Whitney took Furness under her wing, and Furness for her part saw her education extending into graduate student work, with a career in research like Whitney's beyond that, but her family had a more traditional path in mind. In deference to her family's wishes, after graduation in 1891 Furness tried to make a go of the usual late nineteenth-century educated woman's profession of high school teacher, but felt the pull of research too keenly, and 1894 saw her back at Vassar, now as astronomical assistant to Mary Whitney.

When Columbia's Harold Jacoby contacted Whitney for somebody who could reliably interpret astronomical photographic plates, Whitney sent Furness, who in due course was taken on as a graduate student, in spite of the Department of Pure Science's restriction of women students. In 1900, she became the first woman at Columbia to earn a PhD in astronomy with her work cataloguing stars within one degree of the North Pole.

A fresh PhD in her pocket, Furness returned to Vassar, advancing from instructor in 1903 to assistant professor in 1911. From 1910 to 1915 she was effectively running the astronomy department at Vassar as Whitney's health began a precipitous decline, leading to her retirement in 1915. In 1905, she expanded her dissertation's catalogue of sixty-five stars within one degree of the North Pole to a study of 408 stars within two degrees of the North Pole, and in 1913 she co-edited with Whitney *Observations of Variable*

Stars, which included observations of 4,800 variable stars which had resulted from the pair's ten-year observational programme. That experience then fed into her definitive text on the subject, 1915's *Introduction to the Study of Variable Stars*.

She was inducted as a fellow of the Royal Astronomical Society in 1922.

Evelyn Frances Leland (*c.* 1870–*c.* 1930)

Evelyn Leland acted as a human computer at Harvard Observatory from 1889 to 1925, serving in both the Pickering and Shapley eras. Her work centred around identifying variable stars and stars with unusual spectral lines.

Anne Sewell Young (1871–1961)

If Mount Holyoke proved such a regular training ground for future women astronomers in the early twentieth century, much of the credit for that can be laid at the feet of one of that college's most gifted professors and mentors, Anne Sewell Young, who taught at Holyoke from 1898 to 1937 and was a regular author of a popular astronomy column that was probably the inspiration for the direction in scientific communication taken by her famous student, Helen Sawyer Hogg. Young's father was the Princeton astronomer and textbook author Charles A. Young, and so it was not too surprising when young Anne took to astronomy as well, graduating from Carleton College in 1892, then teaching mathematics at Whitman College for several years before returning to Carleton for graduate work, receiving her master's in 1897.

The following year, in 1898, she began her momentous career at Mount Holyoke while pushing ahead with her PhD work at Columbia University, ultimately writing a dissertation on the constellation Perseus that uncovered twice as many stars in the constellation as originally measured. In 1899, she took up a position as director of Holyoke's John Payson Williston Observatory to complement her role as head of the Astronomy Department, which dual roles she fulfilled for the next thirty-eight years. In 1911, she co-founded (with seven other astronomers of whom she was the only woman) the American Association of Variable Star Observers (AAVSO), an organisation of which she was president in 1923, and to which she contributed observations of over 6,500 variable stars.

She was particularly remembered as a teacher, one of those underpinning presences on campus who were known to provide a safe and inspiring place in which to wrestle with large questions. As her former student Helen Sawyer Hogg wrote in her obituary of Young, 'The beauty of the constellations really shone forth as Miss Young taught them to us at night, and objects like the Grand Nebula of Orion, seen through the 8-inch refractor, acquired a mystic radiance which has never dimmed.' Young was pleased to see that one of her former students, Alice Farnsworth, succeeded her at the Williston Observatory, and one of the constant pleasures of her old age was writing letters to the vast array of students she had taught over the previous decades to, from time to time, look up and note what they saw.

Agnes Ermina Wells (1876–1965)

Agnes Wells is today known primarily as a college administrator and women's rights activist on the strength of her near-twenty years as dean of women at Indiana University (1919–1938) and her outspoken views on the relegation of women who desire to work to those jobs too unfulfilling to be of interest to men. Her work trying to improve the lives and prospects of the young women entrusted to her care did take up a significant portion of her time, but after stepping down as dean of women she was able to return to mathematics and astronomy as a professor, and in her early years as dean of women she was somehow also able to complete her PhD (on the subject of radial velocities, i.e. the component of the motion of celestial objects that is towards or away from the Earth) in 1924. Her life after retiring from Indiana University in 1944 was largely filled with her work for the National Woman's Party, which attempted an early version of the Equal Rights Amendment in 1950 that passed the Senate but failed to reach the House.

Margaret Harwood (1885–1979)

When Maria Mitchell died in 1889, her native Nantucket decided to form an association to honour and carry on the work of its most scientifically famous daughter. And so, in 1902 the Maria Mitchell Association was founded, and had among its early priorities the establishment of an observatory in Mitchell's honour that would provide research positions for women astronomers. That institution, the Maria Mitchell Observatory, was founded in 1908 and was guided during the first four decades of its existence by the steady hand of Margaret Harwood (director 1916–1957), who was the first woman to run an independent observatory. She had come to astronomy the way many did in that era, through the gates of Harvard Observatory, which she would visit as a Radcliffe student lodging with the astronomer Arthur Searle, who regularly took her along with him on his visits. She formed friendships with Henrietta Swan Leavitt and Annie Jump Cannon and, upon graduation, naturally fell into an assistant position at the Observatory in which she continued until her Nantucket appointment. Like her near contemporary Alice Farnsworth, she was a specialist in photometric analysis, which uses the intensity of the different wavelengths of light gathered from a star to determine information about the star's luminosity or distance, and which we will look at in more detail in the portrait of Farnsworth below. In 1957, she reluctantly stepped aside from her position at the Maria Mitchell Observatory, being succeeded by Dorrit Hoffleit (whom we shall also meet below). The astronomical world did not, however, forget about her in retirement, naming an asteroid after her in 1960 (7040 Harwood), and in 1961 awarding her the Annie Jump Cannon Prize.

Ida Barney (1886–1982)

Though it was the most famous institution to do so, Harvard was not the only observatory that leaned heavily on the work of underpaid women in the early twentieth century. Yale was in on the practice too, and in the process produced its own crop of superstar computers turned astronomers, of whom few rank as high as Ida Barney. A native of New Haven, Connecticut, she enjoyed the best of her era's higher education, attending Smith College for her bachelor's degree and obtaining her PhD in mathematics from Yale in 1911. She held a number of professorships in mathematics until her astronomical career began in 1922 with her appointment to the staff of the Yale Observatory, tasked with the labour-intensive process of taking star images from photographic plates and computing their positions as celestial coordinates. Her supervisor, Frank Schlesinger, firmly believed that women did not have the mental ability to perform anything but brute force computational work, and so Barney did not receive the encouragement to form theories from her observations. Cut off from theoretical work, Barney worked to improve the procedures and mechanisms involved in computing celestial positions, including introducing more automation and electronic machinery to the process to increase accuracy and reduce eye strain on the observatory's pool of computers. She took over Schlesinger's role of catalogue supervisor upon his retirement in 1941, continuing in that role until 1955. All told, she spent thirty-three professional years at Yale, and in that time personally contributed calculations of the position, proper motion and magnitude of 150,000 stars, her work forming a vastly important part of the eventual Yale Observatory Zone Catalogue's 400,000-star database, concluded the year after Barney's death in 1983.

Hannah Steele Pettit (1886–1961)

Pettit was one of the great solar eclipse hunters of her era, participating in observational expeditions to Colorado in 1918, Honey Lake in 1923, Point Loma in 1930 and Lancaster in 1932, working with her husband, Edison Pettit, to produce high-quality images of eclipse events. She was associated with several different observatories during her career, including Yerkes, Mount Wilson and Sproul.

Julie Marie Vinter Hansen (1890–1960)

Julie Vinter Hansen was her era's gold standard as a calculator of minor planetary and comet orbits, whose accuracy in accounting for orbital perturbations was of an almost legendary order. In an era before access to digital computation, she computed an orbit for the comet Comas-Sola that, by the end of quarter of a century, was still accurate to within one quarter of one minute of arc, to give just one example of the rigour and reliability of her work. She spent her career at the University of Copenhagen, engaged in intensely detailed observations of the complicated dances of the solar system's less heralded objects, the asteroids and comets that are at the whim of not only the Sun, but the more massive planets that lie along their orbital paths. This was classical astronomy which Johannes Kepler might have felt at home with, harnessing the power of more powerful twentieth-century observational equipment to continue a task begun in the deepest past.

More than her era's most iron-clad orbital computer, Hansen was also a distribution nexus of early twentieth-century astronomy, much as Agnes Clerke had been in the previous century. She edited both the *Nordic Astronomy Review* and the important *Circulars of the International Astronomical Union,* which kept astronomers the world over abreast of each other's research. In recognition for her orbital and editorial work, she was admitted to the Royal Astronomical Society in 1931.

Maud Worcester Makemson (1891–1977)

Archaeoastronomy, or the study of the discoveries and methods employed by ancient astronomers, is a field of study we usually associate with our own modern academic sensibilities – how uniquely wise we are, to attempt to reconstruct the tools and knowledge of the non-Western past, unlike our crude Eurocentric forebears. Unfortunately for that particular narrative, there exists the shining figure of Maud Makemson, who during the 1940s and 1950s attempted to reconstruct the astronomical heritage and wisdom of Mayan and Polynesian societies in a series of books including: *The Astronomical Tables of the Maya*; *The Morning Star Rises: An Account of Polynesian Astronomy*; and her translation of the sixteenth-century text, *The Book of the Jaguar Priest*.

Perhaps Makemson's particularly broad interests were a result of her uniquely long route to astronomy. She grew up in New Hampshire, attended – but did not graduate from – Radcliffe, before taking up a career as a teacher and later journalist (as a reporter for the *Phoenix Gazette*), who came to an interest in astronomy in her twenties, and took correspondence courses to fill in her educational gaps while saving money for college by teaching in the southern California public school system. She began attending, depending on the source you believe, either UCLA or UC Berkeley, in her mid-thirties, and received her bachelor's degree in 1925, culminating in her PhD from Berkeley in 1930. In 1932, she began her association with Vassar, where she would remain until her retirement while pursuing various side projects, including her time as an investigator of Mayan astronomical practices on a 1941 Guggenheim Fellowship in ethnoastronomy. After her retirement from Vassar in 1957, she co-wrote a textbook on astrodynamics with Robert M.L. Baker, and advised NASA on its prospective plans for lunar exploration.

Susan Raymond King (1892–1970)

Take a stroll down the wide lanes of the Internet, and in spite of all the vast collected wisdom of the world gathered there, all you will find about Susan Raymond King is a picture of a tombstone. Largely, this is a factor of her research career having been cut short by marriage before it could reach its full potential. Graduating from Smith College in 1913 in mathematics and astronomy, she went on to Harvard (Radcliffe College) to continue her studies in both, receiving her master's in 1919. From 1917 to 1922 she taught astronomy at Smith, until her marriage to a chemist by the name of Harold S. King compelled her to move to his place of work in Nova Scotia. She was not able to resume her astronomical career until her husband took up a job in the Second World War, doing chemical ordnance work for the United States, giving Susan the opportunity to steal some time to do more research at Yale on variable stars. After the war, however, Harold's continued work for the US army meant Susan had to, once again, give up her astronomical career, never to return.

Alice Hall Farnsworth (1893–1960)

Somewhere in between classical astronomy, which measures where celestial objects are, and spectroscopy, which spreads light out to look at the elements present in a star, lies photometry, which is a type of astronomy interested in the magnitude of the light collected from a star across different frequencies. Here, the astronomer uses different filters (called 'passband' filters) to isolate particular frequencies of light and record their intensity. By observing a star through a series of different filters, you can build up the star's 'light curve' which provides the magnitude of light put out by a star across different wavelengths, and thus tells you information about the star's colour, luminosity and distance. Alice Hall Farnsworth was one of the early twentieth century's most significant photometrists, producing among other writings a 1928 compendium of photometric data for 1,550 stars, as well as studies of the properties and effectiveness of microphotometers.

Farnsworth attended Mount Holyoke College as an undergraduate, studying under Anne Sewell Young, whom we met above, though she was not at Holyoke at the same time as our friend Helen Sawyer Hogg, as some online sources claim (Farnsworth graduated in 1916, Hogg did not begin attending Holyoke until 1922). For her graduate work, she moved to the University of Chicago, which ran the Yerkes Observatory (which had begun effective operation in 1897), taking a master's and PhD in quick order in 1917 and 1920. Her dissertation was, characteristically for the work she would go on to do, a study of the technical specifications of the different astronomical instruments at Yerkes, as to their utility for photometry. Most of her subsequent professional career was spent at Mount Holyoke, where she became a full professor in 1937 after seventeen years with the institution. Besides photometry, her interests also included work with sunspots and lunar occultation (what happens when a star disappears behind the dark limb of the Moon).

Pelageya Shajn (née Sanikova) (1894–1956)

We do not know a great deal about Pelageya Shajn, who in 1928 became the first woman to discover a minor planet with the observation of asteroid 1648 Shajna. She was born to a peasant family from the Perm region of Russia, located in the middle Ural mountains. The next we know of her, she is working at the Simeiz Observatory on Mount Koshka (literally, Cat Mountain) in Crimea, and is the wife of Grigory Shajn, who had a number of research interests, including the inevitable stellar spectroscopy, but also the rotation rates and radial velocities of stars. During her career, Pelageya discovered fifteen minor planets, and some 140 variable stars. In 1949, she co-discovered comet 61P/Shajn-Shaldach, a faint comet with a period of 7.49 years, and which made its last closest approach in October of 2022, so if you were planning to try and see it by way of a personal pilgrimage in the memory of Pelageya Shajn, you are going to have to wait until 2029.

Emma T.R. Vyssotsky (née Williams) (1894–1975)

Few individuals took as circuitous a path to a career in astronomy as Emma Vyssotsky. A graduate of Swarthmore College in mathematics, after college she undertook work as a demonstrator at Smith College before making the unique decision to spend some time flexing her mathematical might as an actuary for a life insurance company. After the First World War, she beheld the humanitarian crisis breaking out in Germany, and gave several years of her life working to ensure that the children of Weimar Germany had regular access to food. Returning to the United States, she pushed ahead with her actuarial work for a few more years before making the decision to study for a PhD in astronomy at Radcliffe, a degree she ultimately earned in 1930 for her work on type A stars (white stars with a temperature from 7,500 to 10,000 Kelvins). After the acceptance of her dissertation, she took up a research position at the McCormick Observatory at the University of Virginia. Here she met the Russian astronomer Alexander Vyssotsky, whom she married. The two worked together until Alexander's retirement in 1958, including the co-authorship of a book on stellar motion.

Mary Lea Heger (1897–1983)

Mary Lea Heger's career as a researching astronomer lasted less than a decade, but in that time she pulled off one of the early century's most significant discoveries. When light from a distant star passes through the interstellar medium, certain frequencies of that light are snatched by the atoms within that medium, and used to hop their electrons up to a higher energy level. As a result, the light observed from that star will feature several distinct dark, non-Doppler-shifted bands (because the absorption happened in the intervening medium, and not at or near the moving star), denoting the type of atoms encountered between the observer and the star. Heger was the first to discover such diffuse interstellar bands, and to deduce from them the presence of sodium in the interstellar medium. It was a major result which deepened our knowledge of the cosmic stuff that lies between the stars, but Heger did not have much opportunity to follow up her work, as in 1928 she gave up her astronomical career to raise a family, returning to scientific practice during the Second World War as a calculator for the Manhattan Project, and eventually undertaking the role of scientific historian, organising and categorising the letters, documents and artefacts she found detailing the early history of the Lick Observatory (where her husband was director), work which currently forms the core of the Mary Heger Archives at that institution.

Frances Woodwroth Wright (1897–1989)

In this book, we have not often had a chance to talk about the practical aspect of astronomy as affecting the lives of people who do not happen to be astronomers themselves or to be counted among the Astronomically Curious. Frances Wright stands as a prime example of an individual who took her knowledge of the stars and brought it back down to the Earth for the benefit of people attempting to make their way across its watery surface. She was the author of four different books about marine navigation, one of which, *Basic Marine Navigation* (1944), she published as an extension of her work in the Second World War teaching celestial navigation to military officers. She continued that drive to better educate sailors on the use of astronomy in navigation with *Celestial Navigation* (1969), *Particularized Navigation: How to Prevent Navigational Emergencies* (1973), and *Coastwise Navigation* (1980). Like many women astronomers of the era, she worked as a computer at Harvard Observatory in the Shapley Era (and was in fact a great friend of Cecilia Payne-Gaposchkin), earning her PhD at Radcliffe in 1958. Her non-navigational work included investigations of meteoric particles, spherules (bits of cosmic dust), and methods to calculate galaxy brightness.

Charlotte Emma (Moore) Sitterly (1898–1990)

The scientific career of Charlotte Moore Sitterly breaks into two neat halves: her twenty-five years (1920–1945) at Princeton University working as assistant to Henry Norris Russell, author of one of the early twentieth century's standard astronomical textbooks and part-time villain in the tale of Cecilia Payne-Gaposchkin, and her subsequent twenty-three year stretch (1945–1968) at the Bureau of Standards, leading the effort to obtain and document the atomic spectra for all elements. The child of two teachers who had to tutor and substitute teach to pay her way through Swarthmore College, she more or less fell into working with Henry Norris Russell after failing to obtain a fellowship that would have allowed her to enter graduate school.

Russell and Sitterly were, for quarter of a century, the Odd Couple of American astronomy, with Russell's impetuousness and tendency to hop between different astronomical subjects counterbalanced by Sitterly's compulsion to do work gradually and solidly, with a preference for digging deeply into a subject rather than sampling every plate in the astronomical buffet. The pairing also resulted in one of my favourite stories in astronomy, which tells the tale of how Sitterly came to give up using soft pencils. Russell, it seems, was absolutely mad for soft pencils, and would steal them every chance he got, including those brought to work by Sitterly. Sitterly did what she could to find his treasure troves of soft pencils and return them to their rightful owners, but as for herself decided the only way that she was going to ever get things written down reliably was to switch to a type of pencil that Russell did not hunger after, and so she made the switch to harder leads. I find everything about that story delightful.

Sitterly's main object of research, when not making observations and assembling massive data banks for Russell to use in his own studies, or coordinating the international effort to gather and publish full spectral and energy data for all of the elements in existence, was the Sun. She was an expert in the study of sunspots, and gathered as much information as she could about the spectrum of the Sun across many different wavelengths in and around her official work, and especially the new information on the X-Ray and Ultraviolet regions of the solar spectrum made available by rocket-assisted observations.

For her work on sunspots, and stellar spectra, she was elected to the Royal Astronomical Society in 1949. Her tables of atomic spectra are still in use today.

Vera Fedorovna Gaze (1899–1954)

The St Petersburg-born Vera Fedorovna Gaze was a specialist in the field of emission nebulae. Nebulae are essentially just massive clouds of dust and gas, which can stretch for dozens upon dozens of light years and yet have incredibly low densities. Sometimes nebulae that are close to a star will be hit with ultraviolet light coming from that star, which will cause the nebula's gases to ionise, and give off light over a variety of wavelengths, often with striking multicoloured results. These nebulae, called 'emission nebulae', were Gaze's particular area of expertise, and over the course of her career, she discovered something on the order of 150 of them before writing an important text on nebulae formation in 1952, *Some Results on the Study of Diffuse Gaseous Nebulae and their Relation to Cosmogony*. In 1977, the minor planet 2388 Gase was named in her honour by N.S. Chernyk.

Henrietta Hill Swope (1902–1982)

When Henrietta Swan Leavitt elucidated the relationship between luminosity and period in Cepheid variable stars in 1912, she created a new field of observation and study for the generations that succeeded her; one of the many to directly benefit from her work was Henrietta Hill Swope, whose famous family (her uncle was a member of the Algonquin Round Table that included Dorothy Parker, Tallulah Bankhead and Herman Mankiewicz) and elevated upbringing little suggested the brass tacks, rigorous astronomer she would become.

Her education was unlike most of those we shall come to known in the twentieth century – she took an associate's degree in mathematics from Barnard College in 1925, and then a master's degree from Radcliffe in 1928 while working with Harvard Observatory director Harlow Shapley (who had taken over the HO two years after the retirement of Pickering in 1919), and that was it. She stood more in the nineteenth-century mould of a passionate and talented worker who began at the labour-intensive bottom of the profession, identifying variable stars in the Harvard Observatory trenches alongside future superstar Cecilia Payne-Gaposchkin, and using her technical expertise to build a reputation that would gain her employment usually open only to those possessing PhDs. From 1952 to her retirement in 1968, she worked at the Carnegie Institute of Washington, alongside Andromeda Galaxy researcher Walder Baade, with whom she used Andromeda Cepheids to produce ever more accurate estimates of the distance between the Milky Way and Andromeda galaxies, obtaining a distance modulus of 24.25 that compares favourably with the currently accepted value of 24.4.

Just before retiring, she donated $650,000 in securities to the Carnegie Institute, for the purpose of augmenting the world's astronomical instrumentation in the Southern Hemisphere, money which ultimately went to the erection of the 40-inch Swope Telescope in Chile in 1971.

Wilhelmina Iwanowska (1905–1999)

Nicolaus Copernicus. Johannes Hevelius. Wilhelmina Iwanowska. For most people, these three represent the working summary of the history of Polish astronomical research from the sixteenth through to the early twentieth centuries – a couple of profound mountain peaks placed in otherwise unremarkable and flat historical valleys. With Iwanowska, Poland's first professional astrophysicist, that trend of intellectual feast and famine began to steady itself, largely due to the institutions she helped create and develop. She began her career at the University of Stefan Batory, located in the Lithuanian capital of Vilnius/Wilno, where she received her bachelor's and master's degrees in mathematics in the late 1920s. Over the course of the 1930s, she was responsible for popularising spectroscopic research in Poland, the methods of which she had learned during a 1934 internship at the Stockholm Observatory, pushing Polish astronomy away from its distinguished past in positional astronomy, and bringing it in line with the methods that had already been in steady use for decades in Western Europe. In 1937, she demonstrated the power of the new technique in her post-doctoral dissertation characterising stellar super-giants by their spectra.

The war years were, to put it mildly, a difficult time for the Polish people, and in 1945 the scientific community was faced with the task of reassembling itself. Iwanowska was a key player in this process, co-founding the Nicolaus Copernicus University in 1945 as a university of refuge for some 200 academics who had formerly worked at the University of Stefan Batory. The new university boasted departments of astronomy and astrophysics, and it was here in 1946 that Iwanowska became her country's first professor of astrophysics. She remained at NCU for the next fifty-four years, supervising nineteen doctoral students (of whom eight went on to become professors themselves), working until her death in 1999 at the age of 93.

Helen Dodson Prince (1905–2002)

Helen Dodson Prince was one of the twentieth century's outstanding researchers in the field of solar flares. Other astronomers might drift from cosmic flower to flower, but Dodson Prince was laser focused on elucidating solar flares and only solar flares over the course of a forty-six-year long career, a model of concentration and devotion almost unimaginable to our frayed and worn modern attention spans. Originally interested in mathematics, which was her major at Goucher College, and the arena of her first job – as a statistician for the Maryland Board of Education – her growing interest in astronomy eventually took her to the University of Michigan, where she earned her master's degree (1932) and PhD (1933, for a spectrographic study of 25-Orionis) before travelling to France for two summers to work at the Paris Observatory at Meudon, where her interest in solar flares took flight.

Like many scientists, the arrival of the Second World War caused Dodson Prince to put her astronomical career on hold, while she employed her training in physics and mathematics at MIT's Radiation Laboratory (where Henrietta Swope also worked) in radar research. The end of the war allowed Dodson Prince to return to her solar research, where until her retirement in 1979 she devoted herself to the writing of some 130 papers on the subject of solar flares, the development of the Comprehensive Flare Index (CFI) in the late 1960s to categorise the magnitude of a solar flare event based on five observational categories, and the publication of a magisterial report in 1975 summing up the results of fifteen years of solar flare observations at the McMath/Hulbert Observatory at the University of Michigan.

Ellen Dorrit Hoffleit (1907–2007)

Some August night in 1919, a young Dorrit Hoffleit was taken outside at night by her mother, the daughter of a Prussian physics professor, to look at something incredible: the annual display of the Perseid meteor shower. An astronomical event caused by the comet Swift-Tuttle which reaches its crescendo of brilliance with stunning regularity around 12 August each year, the sight inspired Hoffleit to learn more about the night sky. Born in Alabama, she grew up in New Castle, Pennsylvania, and attended Radcliffe University, from which institution she earned her bachelor's degree (1928), master's degree (1932), and PhD (1938). After receiving her bachelor's degree, she began working as a research assistant at the

Mecca of early twentieth-century American astronomy, the Harvard College Observatory, which had just seen the end of the esteemed Pickering Era (1877–1919) and was well into the Harlow Shapley years (1921–1952) (fun fact: there has never been a woman director of the Harvard Observatory, in spite of the central role women have played throughout the history of that institution).

At Harvard, Hoffleit worked with photographic plates, much as a string of world-famous woman astronomers had done before her, and discovered approximately 1,000 Southern sky variable stars. Her work was of superior quality, and noting this Shapley encouraged her to work towards a PhD. Her dissertation, *On the Spectroscopic Determination of Absolute Magnitudes, With Application to the Southern Stars of Types Later than A*, was deemed the best original work that year. With a reputation for rigorous work, and well-placed mentors, Hoffleit was well set to pursue a career in astronomy when the Second World War broke out. She decided to donate her time and talent to the war effort, working as a mathematician for the Aberdeen Proving Grounds Ballistic Research Laboratory in Maryland, and later as a scientist on the project to import and absorb the rocketry know-how brought to the United States by Wernher von Braun and the members of his V-2 project.

It was not until 1948 that Hoffleit returned to astronomy, resuming work at Harvard Observatory, which went well until Shapley retired in 1952 and the relatively short-lived directorship of Donald Menzel got underway (1952–53, 1954–66). Menzel had his own ideas about what the Harvard Observatory should be doing, and Hoffleit's approach struck him as belonging to an earlier and obsolete era, and he told her as much, prompting her to leave Harvard in 1956 in favour of Yale and the Maria Mitchell Observatory, remaining at the Yale Observatory until 1975, and directing the Maria Mitchell Observatory until 1978. Her research interests at Harvard and Yale were broad, and included variable stars, proper motion, stellar spectra, meteors, galactic structures, quasar variability, and the compilation of two massive reference works, the *Bright Star Catalogue*, which gathered information on over 9,000 of the sky's brightest stars and continued the tradition of her Yale predecessor Ida Barney, and *The General Catalogue of Trigonometric Stellar Parallaxes*, which gathered distance measurements to over 8,000 stars which proved fundamental to better understanding the motion and evolution of the Milky Way. Though retired from her observatory positions by 1978, she continued her research into her nineties, before passing away from cancer at the age of 100.

Nüzhet Gökdoğan (1910–2003)

When Nüzhet Gökdoğan joined the astronomy faculty of Istanbul University in 1934, she was the only individual actually from Turkey working there. The department at the time was centred on the work of Erwin Freundlich, whom we shall meet later in the career of Paris Pişmiş, and boasted a staff in the single digits which it would be Gökdoğan's job to expand and develop over the next nearly half a century. At the age of 18, she had won a scholarship to study in France, earning a bachelor's degree in mathematics from the University of Lyon in 1932, followed by a Diplôme d'Études Supérieures from the University of Paris, and an internship at the Paris Observatory.

The pentalingual Gökdoğan (she spoke German, English, Greek, French and Turkish fluently) could have built her career anywhere after that point but chose to return to Turkey, where she was turned down for a position at Kandilli Observatory, explicitly on account of her gender. Instead she took a position at Istanbul University, where she became the first woman member on the science faculty. Here she completed her PhD in 1937 on interstellar dark matter surrounding the Sun, which leads us to the minor historical controversy of 'Who was the first woman from Istanbul University to earn a PhD?' Gökdoğan's thesis is registered as the first completed at the university's faculty of science by anybody, while her near exact contemporary Paris Pişmiş is almost universally tagged with the epithet 'First Woman To Earn a PhD at Istanbul University'. If anyone out there has access to the university's records and a spirit of investigation, please check it out, and let me know what you find so we can settle this once and for all!

Over the years to come, Gökdoğan was a driving force behind the scientific evolution of Turkey, including: founding the Turkish Mathematical Society in 1948 and Turkish Astronomical Society in 1954; organising a number of international symposiums in Turkey; establishing cooperative programmes with observatories in Italy, Switzerland and France; becoming Dean of the science faculty at Istanbul University in 1954 (the first woman in Turkey to hold such a position); and leading the astronomy department as its chair from 1958 to 1980, during which time the staff grew from five to eighteen members, and she personally oversaw eleven PhD dissertations.

Paris Pişmiş (1911–1999)

And now we get down to Pişmiş. Over the course of her long life, she travelled the globe, making history more or less wherever she went. Born to Armenian parents in Istanbul, she attended the Istanbul University and became (apparently) the first woman to earn a PhD in science from that institution in 1937, following up that accomplishment with a trip to the United States to study at Harvard in 1938 upon the advice of her advisor Erwin Freundlich, famous for his proposal of certain astronomical observations that could establish the veracity of relativity theory. While at Harvard, she worked as an assistant at the Harvard Observatory at a rate of 50 cents per hour, and it was here, in 1940, that she met a visiting mathematician from Mexico by the name of Felix Recillas. The two fell in love, and married in 1941.

In 1942, the Tonantzintla Observatory was established just in time for the arrival of the newlywed couple, and Pişmiş was brought on as the observatory's primary astronomer, making her Mexico's first professional astronomer. After some time in Princeton and at the Yerkes Observatory, she returned to Mexico in 1948, teaching at the National Autonomous University of Mexico for the next half century, while at the same time carrying out research in multiple fields, including: the study of moderately active galaxies; the formation of spiral galaxies; the motion of ionised nebulae; and open star clusters (groups of a couple of thousand stars formed from the same molecular cloud, of which Pişmiş discovered twenty). Just before her death, she published an autobiography, *Reminiscences in the Life of Paris Pişmiş: A Woman Astronomer*, which I have never been able to find a copy of outside an institution, so if you come across one, let me know.

Liisi Oterma (1915–2001)

Liisi Oterma is not only Finland's most well-known woman astronomer but perhaps its most well-known woman scientist generally. Eventually fluent in at least eight languages (and rumoured to be as many as eleven), she chose mathematics and astronomy as her focus at the University of Turku in Finland when she found out she could not study Sanskrit there by way of pursuing her love of foreign tongues. The astronomy department at Turku was relatively young when she came to it, having been established only in 1924, and consisted almost entirely of one person, Yrjö Väisälä, and his research assistant. Väisälä was something of an optics nut, and used a method of exposure overlapping to try and catch minor planets in the act of moving against a static background. Oterma aided these efforts, discovering, by the reckoning of the Minor Planet Center, fifty-four minor planets between 1938 and 1953, though other reckonings place her total at as much as 200 over her career. In 1938, Väisälä named asteroid 1529 Oterma (part of the Hilda family of asteroids that are regularly acted upon by Jupiter's gravity) in her honour.

In 1955, her dissertation on telescope optics was accepted, making her the first Finnish woman ever to earn a PhD in astronomy and earning her the 'Woman of the Year' award from the Finnish Alliance of Business and Professional Women the following year. In 1965, she became a full professor at Turku, and upon Väisälä's death in 1971 took up his position as director of astronomical-optical research and director of the Tuorla Observatory. She retired from the University in 1978.

Jocelyn Ruth Gill (1916–1984)

In a fully just universe, there would be much more of Jocelyn Gill in the annals of astronomy than the few fragmentary online pages currently dedicated to just barely remembering her. In that universe, she would not have been diagnosed with multiple sclerosis in 1966 – at the height of her career as chief of NASA's in-flight science for the manned space programme – and she would not have been compelled to retire at the hands of that disease in 1973 at the age of 57, to head bravely into her remaining eleven years of life.

That not being the universe we happen to inhabit, however, we have to rest content with the magnificent things Gill did accomplish during her life, like her glittering record as a student, graduating from Wellesley, then receiving her master's from the University of Chicago for her work at the Yerkes Observatory, and finally her PhD from Yale. Or we can celebrate her record of innovation at the dawn of the American space programme, as a designer of the experiments performed on board the first three great waves of NASA manned flight – the Mercury, Gemini and Apollo programmes. Or that her scientific spirit did not stop at designing experiments, but included carrying them out, including her personal trip aboard a flight 8 miles above the surface of the Earth to record the Sun's corona during a 1963 eclipse. We can certainly wish that her post-graduate school career had extended further than the fourteen years it did, but it is hard to argue against the fact that, in those fourteen years, she accomplished more for women as scientific administrators, for herself as a scientific investigator of the as-yet unknown phenomena governing the universe beyond our Earth, and for the visibility of other people suffering from the disease that curtailed her career, than any of us would have any right to expect in a career three times as long.

Alla Genrikhovna Massevich (1918–2008)

The Georgian-born Alla Massevich wore many hats during her astronomical career in the USSR, from her role erecting something on the order of sixty observation stations within the span of a half a year to track the motions of Sputnik across the sky, to her critical role in establishing planetary geodesy (the study of the size, shape and gravitational fields of planets) as an important branch of astronomical study in Russia, to her diplomatic role as an international spokesperson and liaison for Soviet space science. And that tremendous career all began, in a manner almost too charming to stand, with a series of letters between a science-loving young girl and a generous and encouraging science communicator. Yakov Perelman (1882–1942) was a beloved author of a series of books attempting to make science and mathematics exciting and approachable for young people and passionate amateurs, with titles like *Mathematics Can Be Fun* and *Geometry for Entertainment* (most of which, by the by, are still very much available in reprints). In her early teens, Massevich came across one of these books and was so enchanted by it that she wrote Perelman a letter with some of the questions she had about the book and, as luck would have it, he wrote back, part of a correspondence stretching over years that encouraged and inspired the young girl to take up a career in science, and to pass on the favour as an adult by taking the time to write regularly to young people interested in astronomy and space travel.

She earned her bachelor's degree in mathematics and physics just precisely in time for the Nazi invasion of the USSR to disrupt her graduate studies, compelling her and her husband to flee to the relative safety of the south-eastern part of the country, where she completed her studies and received her doctorate in 1946. By 1952, she was the deputy chairman of the Astronomical Council of the Academy of Sciences of the USSR, and in 1957 she was placed in charge of the project to optically track Sputnik and future Soviet satellites, to which end she pulled off the mammoth administrative feat of erecting and coordinating over five dozen observation stations across the USSR. Over the 1960s, she carried on her original research in planetary geodesy while emerging as one of the most internationally popular and respected faces of the Soviet space exploration movement, serving as an administrator for various organisations aimed at encouraging international scientific relations during the politically tense years of the Khrushchev-Brezhnev era. After some time in the 1970s teaching geodesy at the Moscow Institute of Geodesy and Cartography, she ascended to the position of Chairman of the Astrosoviets in 1987, which sounds like an alternative title for a science fiction galactic emperor, but which in fact placed her in charge of the Astronomical Council where she had been serving in the capacity of Deputy Chairman during the previous thirty-five years.

Anne Barbara Underhill (1920–2003)

Anne Underhill will forever be associated with the early-type stars which her life was dedicated to elucidating. In astronomy, the word 'early' is a stand-in for the word 'hot' and tends to refer to stars of type O and B – highly luminous blue-coloured stars at high temperatures with brief life spans that were largely blanketed in mystery when Underhill began her career in the mid- to late 1940s. Most of her career before 1962 was spent in her native Canada, where she laboriously harnessed the power of the first generations of computers to begin developing the mathematical models governing early-type stars. In 1962, she was offered a full professorship at Utrecht University, where she worked for the next eight years, writing *Early Type Stars* (1966) which became the standard text on the subject, and continuing her studies on O and B stars unencumbered by some of the more blatantly sexist practices she had borne while at the Dominion Astrophysical Observatory. The last fifteen years of her career were spent at the NASA Goddard Space Flight Center, where she was a guiding force behind the International Ultraviolet Explorer Satellite, which launched in 1978, and the data from which would prove critical for the work of another great woman astronomer we shall meet shortly, Erika Böhm-Vitense.

Alenush Terian (1921–2011)

Terian's name today is inseparable from the phrase 'Mother of Modern Iranian Astronomy' on the strength of her co-founding of the solar observatory at the University of Tehran, her status as the first woman to become a full professor of physics in Iran, and her three decades of work in solar physics, often under severe limitations of funding and resources, including ten years as the chief of solar physics research at the University of Tehran. She was the child of Armenian parents who were involved in literature and the theatre, and who were willing to finance their child's graduate-level scientific studies in Paris after her graduation from a Zoroastrian high school in Isfahan and reception of her bachelor's degree from the University of Tehran. In Paris, Terian received her doctorate in atmospheric physics from the Sorbonne in 1956. She was then offered the chance to continue teaching at the Sorbonne, with all of the access to equipment and resources that implied, but decided instead to return to Iran to help build up the astronomy community there, joining the University of Tehran as a thermodynamics associate professor in 1956, and achieving the historical status of full professor in 1964. She retired at last in 1979, and was hailed as an Iranian hero by the government upon the occasion of her ninetieth birthday in 2011.

Erika Böhm-Vitense (1923–2017)

There is a reason that every problem you get in high school physics tells you to 'ignore air resistance', and that is that fluids like the atmosphere make everything an order of magnitude more difficult to figure out the behaviour of. And if a gentle breeze is enough to raise computational hackles, how much more so the super-heated gases in motion above the photosphere of a star. The task of unlocking the motion and behaviour of the different layers of a typical sun fell largely in the mid-twentieth century to a woman who was not offered a steadily paying academic position until seventeen years after she completed her groundbreaking PhD, Erika Böhm-Vitense.

Böhm-Vitense married the astrophysicist Karl-Heinz Böhm in 1953, and thereafter during their joint careers in Germany, anti-nepotism laws kept her in a series of temporary positions at the universities where Karl-Heinz worked, even as her work was contributing fundamentally to our understanding of stellar physics. The same year she got married, she published a paper on convection in the Sun's atmosphere that has for decades served as a starting point for models of stellar atmospheric motion. That paper applied Mixing-Length Theory, which had been developed by aerodynamicist Ludwig Prandtl in the early twentieth century to describe what happens when highly turbulent packets of fluid enter a mass of ambient fluid and distribute their momentum until they are essentially indistinguishable from that ambient fluid. While Prandtl originally developed his theory for solving problems related to aerodynamic flow, Böhm-Vitense thought to apply this concept to the violent convective events of stellar fluids – with great success.

In 1968, Erika and Karl-Heinz moved to the University of Washington, where they could both have stable full-time work, as well as access to an on-campus computer. She remained there for the rest of her career, including her important work in the 1980s using the results of the International

Ultraviolet Explorer satellite to begin unlocking the secrets of the stellar chromosphere (the layer of the Sun between the photosphere, or surface, and the transition region, which lies just beneath the corona) for different star types, and the writing of *Introduction to Stellar Astrophysics*, a three-volume series which served in the role of astronomical bible to two full generations of aspiring stellar researchers.

Ye Shuhua (b. 1927)

The 1920s were not a great time to be born in China. It ensured that one's youth was spent in fear of invasion, adolescence in the grip of Civil War, early adulthood in the throes of food crises and political persecution, and maturity in the crosshairs of the Cultural Revolution. Ye Shuhua had to somehow put together a scientific career in and around the titanic world events snatching at her from all sides, and that she emerged from it all as one of her era's most honoured and respected scientific minds is a testament to her rigour as an investigator and perseverance as a human. Her family had to move three times during the Second World War to escape the oncoming Japanese invasion, and when it was all over, and Ye could attend university, she found herself so hemmed in by the expectations of her family that, in a world so uncertain as that of post-war China, she would turn her brain to a practical profession like medicine, instead of the literature she had set her heart on.

Fortunately, a compromise was reached, and Ye was permitted to study mathematics, which just so happened to share a department with astronomy at Sun Yat-sen University, and by the end of her undergraduate years, Ye had switched her major to astronomy. By 1951, Ye and her husband were in Shanghai, where she worked at the Zikawei Observatory until in 1958 her country needed her to correct one of the most pronounced blemishes on its scientific reputation – its last-place standing in the measurement of Universal Time.

So, what's that?

It was something we did not really need before the advent of rail transport, back when the fastest travel you could manage was dependent on how good your horse was feeling that day, and the furthest you could go was by and large at the same longitude as the place you came from, and so had pretty much the same sense of time. Railroads messed all that up – now the town to which you travelled had a slightly different sense of what time it was from the time at the departure point where you had set your watch, and what was desperately needed was for somebody to come in and say, 'Look, this is when noon is, deal with it.' And so, Greenwich Mean Time was born in 1847. If you lived in England, it meant your noon was when the Sun was directly over Greenwich. By 1884, with the proliferation of reasonably rapid international travel, proposals were underway to expand Greenwich to the rest of the world in a first attempt at creating a universal time standard. Greenwich, however, is not the universe, and eventually it was realised that a more cosmic standard of universal time might be nice, one which was based on Earth's rotation about an angle in an internationally agreed upon celestial reference frame grounded upon a highly stable baseline of very distant stars. If two countries employ the universal time system, they should theoretically be within microseconds of one another in relating the time at which an event happened, making a big difference in international scientific experiments or astronomical observation reporting.

In 1958, China was not good at this, and it was Ye's job to fix it using her skills in mathematics and astronomy. This she did, and within five years, from being the world's worst country in the computation of universal time, it was the second best, and Ye was regarded as a national scientific hero for that feat. Unfortunately, the 1960s were just about the worst possible time to be a scientific hero in China, as in 1966 the Cultural Revolution of Mao Zedong (also known as Chairman Mao) sprung archly to life, seeking to purge the nation of its decadent intelligentsia. Ye was included in this medieval lurch, and from being a national hero she found herself, within a space of years, persecuted and compelled to labour as a house painter to purge her intellectualism. For ten years she was kept from her work while China earnestly waited for Mao to die, which he finally managed to do in 1976. That death, and the collapse of the Cultural Revolution that precipitated the crumbling of the Gang of Four within a month of it, cleared the way for Ye's return to research. By 1981, she had risen to the position of director of the Shanghai Astronomical Observatory (the facility she had been working at back when it was called the Zikawei Observatory), a job she retained until 1993, where she continued her work in tightening how China measures space and time, through the development of very-long-baseline interferometry and satellite laser ranging projects.

In 2021, she achieved a moment of cross-generational fame when a speech on gender equality she gave at the age of 95 rocketed to half a billion internet views thus giving the woman who had overcome war and persecution to make her country a scientific example to the world in her chosen field very nearly the same level of fame as a social media influencer who specialises in opening boxes online. It is a strange world.

Sandra Faber (née Moore) (b. 1944)

Unlike many astronomical specialists, who dig very deep holes, settle into the intense contemplation of their dim surroundings, and like it just fine there, Sandra Faber's career has intersected a bursting jamboree of the twentieth century's most significant astronomical projects across a wide range of research fields. In 1972, she became the first woman on the staff at the Lick Observatory (founded 1888) and was soon making a fundamental contribution to the world of astronomy in the form of the Faber-Jackson relation, published in 1976. This relation stemmed from Faber's observation that a correspondence existed between the brightness of a galaxy and the orbital speed of the stars within it. As in the case of Swan Leavitt above, this provided a magnificently powerful tool for estimating the distance to galaxies. By using spectrometry and Doppler shift to determine the velocities of the stars in a galaxy, you can, with the Faber-Jackson relation, determine the actual brightness of that galaxy, which, when compared against the apparent brightness of that galaxy to our observational equipment, provides a means of determining how far away that galaxy probably is. It is a universal measuring stick that does not require Cepheid stars, and it is very fitting that it came from an individual who received her PhD from Harvard University, the very institution where Swan Leavitt devised our first yardstick some six decades previously.

From that triumph, Faber went on to study the emerging subject of dark matter, which Vera Rubin had been developing in the 1970s and early 1980s, and which Faber contributed to with a 1983 paper advocating for a 'cold' dark matter theory, in which the newly proposed form of matter was made up of as-yet unknown slow-moving particles, over a 'hot' theory, in which it is made up of high velocity neutrinos. She then took the idea of dark matter and worked it into theories about galaxy formation and separation that continue to resonate four decades later.

And was she done there? No, for in the mid-1980s she took another shift in direction, and devoted herself to the task of advocating for the revolutionary design of the new Keck telescope, and overseeing its first observational instrument. In 1990, she led the team which discovered the spherical aberration problem hindering the Hubble Space Telescope, and served as the principal investigator for the Nuker Team, which used that rehabilitated scope to plumb for supermassive black holes at the centre of galaxies (a search, we will see, that eventually led to a Nobel Prize for Andrea Ghez), *and* she was one of the creators of the CANDELS project, which sought to employ Hubble to investigate light produced during the

'Cosmic Dawn' (a period covering the first billion years after the Big Bang) and the 'Cosmic High Noon' (an era occurring 2–4 billion years after the Big Bang which saw galaxies dramatically make their presence known). In short, Faber has done a bit of everything in pushing our universe to better know itself, and the universe, I have it on good authority, is grateful.

Judith Gamora Cohen (b. 1946)

Cohen's contributions to astronomy are spread roughly across two massive branches. The first, her research career, has spawned 200 scientific papers, and given us deep insights into the old, metal-poor objects populating our galaxy's outer halo; a four-year survey of the redshift of 1,500 galaxies; analysis of the shifting content of the light elements carbon, nitrogen, oxygen, sodium, aluminium and magnesium in stars found in globular clusters; and an intense study of magnesium isotopes in dwarf stars that shed light on the dynamics and relative importance of different nuclear chains at different times in a star's life. The second branch consists of her contributions towards the tools of astronomy, and particularly her role as a co-designer (with Jerry Nelson) of the Keck telescope (a twin telescope system composed of two 300-ton telescopes with 10m diameter primary mirrors, each consisting of thirty-six hexagonal sections, the largest in the world at the time of their completion and today the third and fourth largest in operation) and of one its most important light instruments, the Low Resolution Imaging Spectrograph that she employed in her redshift survey.

Lyudmila Zhuravleva (b. 1946) and Lyudmila Karachkina (b. 1948)

Sometimes, history really just does not want you to remember things. In the case of Zhuravleva and Karachkina, it has thrown at us two individuals named Lyudmila, who both worked at the Crimean Astrophysical Observatory, both as planetary astronomers hailing from Ukraine, and both engaged in the task of discovering minor planets, and cast us the challenge of not getting them mixed up ALL THE TIME. According to the Minor Planet Center, Zhuravleva discovered or co-discovered 213 minor planets over the years 1972–1992 while Karachkina discovered 130 from 1980 to 1990. That places them at #70 and #105 respectively on the list of all-time minor planet hunters, which does not sound too impressive until you realise that the frontrunners on that list are composed of astronomical survey projects like Spacewatch (which has racked up 178,000 minor planets) rather than individuals, and that most of the individuals who rank higher than them had careers that spanned into the late 1990s and early 2000s, and benefited from technology not available at the Crimean Astrophysical Observatory where they carried out most of their work. In 1982, Karachnika named mainline asteroid 26087 Zhuravleva in honour of her co-worker, while in 1990, asteroid 8019 Karachkina was named in Karachnika's honour, though oddly not by Zhuravleva.

Hashima Hasan

Hashima Hasan knew that she wanted to explore space from the day her grandmother took her outside to observe the passage of Sputnik overhead. When, twelve years later, NASA landed a man on the Moon, she made it her goal to attend Oxford University and ultimately work for NASA herself. In India, she largely taught herself the necessary mathematics and physics to overcome the patchwork education offered to most girls at the time, and earn her way to the university. Through hard work and a drive to prove herself the equal of the men around her, Hasan won a series of scholarships and awards, culminating in her 1976 Oxford doctorate. She set out on a promising academic career as a physicist in India, which was suddenly interrupted by her family's insistence that she take a husband. She was placed in a room with a man of their choice, given two hours to get to know him, and then asked at the end of those two hours whether she would agree to take him or not. She did, and thereupon had to leave her budding academic career to follow him to his work in the United States.

That could have been the end of the story, but Hasan was determined not to let her knowledge base atrophy. She did what research she could without an official position, and kept up on the important literature of her field, until a change in her residency status removed the existing restrictions on her employment status, and she was able, in 1985, to make her dream of working for NASA come true at last. There she worked on developing software that could simulate the optics of the Hubble Space Telescope. After the Hubble's initial pictures turned out disturbingly blurry, it fell to Hasan to analyse the problem and create workarounds that would produce the best possible data under the given restrictions until a permanent fix could be devised and implemented. In 1994, to give herself more experience in the management side of scientific research, she took up a position as Visiting Senior Scientist to the Ultraviolet, Visible, and Gravitational Astrophysics programme, which allowed her to become a Program Scientist or Deputy Program Scientist for some of NASA's most exciting work of the last quarter of a century, including the Hubble Space Telescope, the Keck Observatory, NuSTAR (the Nuclear Spectroscopic Telescope Array which measures high-energy X-rays from astronomical objects, allowing expanded spectroscopic analysis), and the newest superstar on the block, the James Webb Space Telescope.

Adriana Ocampo (b. 1955)

'As soon as I landed in the USA, I asked, "Where is NASA?"'

Born in Colombia, and raised in Argentina before emigrating to the United States at the age of 14, planetary geologist Adriana Ocampo was a space enthusiast from childhood, creating spaceships out of pots and pans and dressing up her favourite doll as an astronaut. Upon arriving in the United States, she started taking whatever courses would move her closer to her goal of space study, and when after her junior year she had a chance to first volunteer and then work at the Jet Propulsion Laboratory, she grabbed it, continuing her work at JPL while obtaining her bachelor's degree in geology and her master's in planetary geology.

During the 1980s, she was primarily engaged in planet imaging missions, including directing imaging of Mars's moons Phobos and Deimos during the Viking mission (1976–1982), data which then informed the Soviet Union's 1988 Phobos mission. World fame came to her in 1991 for her role in the discovery of the Chicxulub crater, the site of the meteor impact 66 million years ago that was probably responsible for the extinction of the dinosaurs. Beginning in 1989, Ocampo and her team employed satellite technology to find water sources on the Yucatan peninsula, and discovered a series of cenotes, or sinkholes, distributed along an arc. Digging deeper into petrochemical records, they found earlier but largely ignored work suggesting the presence of a roughly circular crater on the peninsula that corresponded with the cenotes. Ocampo's group published their findings in a 1991 *Nature* article that proposed the Yucatan crater as the site of the dinosaur-ending impact hypothesised by Luis and Walter Alvarez a decade previously.

In the following years, Ocampo continued her planetary research, as part of the *Galileo* team analysing Europa, the moon of Jupiter which ranks on everybody's top five list of Favourite Moons for its subsurface ocean, weak magnetic field, possible presence of organic compounds, and partially oxygenated atmosphere. She was also head of strategic planning for the 2011 Juno mission which continues orbiting and observing Jupiter to this day.

Claudie Andre-Deshays Haigneré (b. 1957)

Haigneré is not merely one of the most educated people ever to travel to space, but perhaps ever to exist. She holds certificates or degrees in biology, sports medicine, aviation medicine, space medicine, rheumatology, and biomechanics, and holds doctorates in rheumatology and neuroscience. Over the course of the late 1980s, she worked simultaneously at the Rheumatology Clinic and Sport Traumatology departments at the Cochin Hospital in Paris, at the Neurosensory Physiology Laboratory of the CNRS (*Centre National de la Recherche Scientifique*), *and* after her 1985 selection as the only woman admitted to the French space programme, *Centre National d'études Spatiales* (CNES) among 10,000 applicants, she was also preparing herself for her first space missions.

On 17 August 1996, she became the first French woman in space aboard the Soyuz TM-24 flight, where she spent sixteen days on the Mir space station conducting medical and biological experiments. For her next and last space flight, as commander of the Soyuz TM-33 mission in 2001, she became the first woman to visit the International Space Station, which had just received its first long-term inhabitants in November of 2000. Upon her return, Haigneré entered government service as France's Minister for Research and New Technologies from 2002 to 2004, and Minister for European Affairs from 2004 to 2005, before turning to yet a new field of accomplishment and becoming the founding CEO of Universcience, an educational venture that merged the Cité de Sciences and the Palais de la Découverte into one spectacular science museum. After stepping down as CEO in 2015, she returned to space travel as advisor to the director general of the European Space Agency.

Svetlana Savitskaya (b. 1948) and Elena Kondakova (b. 1957)

We have already read how, after making Valentina Tereshkova the first woman in space in 1963, the Soviet Union backed resolutely away from future missions involving women in space travel, grounding Tereshkova and not seriously entertaining the prospect of further women cosmonauts for two decades. When the second woman cosmonaut (and second woman in space), the high-altitude parachutist, test pilot, and record breaking aerobaticist, Svetlana Savitskaya, did make her historic flight in 1982 aboard Soyuz T-7, it seemed as though the long freeze was over. In 1984, she was a crew member of Soyuz T-12 and in that capacity became the first woman ever to perform a spacewalk, on 25 July 1984, preceding American astronaut Kathryn Sullivan by a little under three months. What was supposed to follow that milestone was an exciting all-woman Soyuz crew, led by Savitskaya, to the Soviet space station Salyut 7 to commemorate International Women's Day. Problems with Salyut 7 ultimately led to the cancellation of that mission, and in 1986 Savitskaya gave birth to her son, thereby ending her career as a cosmonaut.

Rather than picking up the women's space programme with the two crew members who were supposed to fly on the Women's Day mission, Yekaterina Ivanovna and Yelena Dobrokvashina, the Soviet Union decided to ditch the whole effort, and the next Russian woman in space would not come until after its fall. Elena Kondakova was a crew member aboard the Soyuz TM-20 mission that launched in October of 1994, destined for the Mir space station. Kondakova remained aboard Mir for 169 days, making her the first woman to undertake a long-term spaceflight (the second, Shannon Lucid, launched in March 1996 for her eventual 188-day mission in space). Her second and last mission was aboard the NASA shuttle *Atlantis* on STS-84, which was one of a series of Mir-docking expeditions that launched in 1997.

Care to take a guess as to when the next Russian woman in space launched? Five years later? Ten years later? Fifteen? Incredibly, it was seventeen years later, in 2014, when Yelena Serova was a member of the Soyuz TMA-14M mission to the International Space Station, her first and only flight. At the time of this writing in 2022, engineer Anna Kikina, is awaiting her flight in September to become the fifth Russian woman ever in space. A cynical observer might conclude from all of this that the Soviet Union, and subsequently Russia, has been very good at putting women into space when there is a propaganda value associated with the event – a First that it can hold over the United States, or an ideological point to be made – but that without those motivations it has had profound difficulties summoning the administrative willpower to make women's spaceflight an integral and regular part of its programme over the last six decades. To put it in stark numbers, of the 339 individuals the United States has put into space, fifty have been women, who represent 14.7 per cent of the total, a record which is abysmal but is positively towering next to the Soviet Union/Russia, which has put 121 individuals in space, of whom only four have been women, representing 3 per cent of the total.

Tereshkova, Savitskaya and Kondakova were all astounding individuals of great bravery and accomplishment across multiple fields. Their example deserves following, and here's to the day it happens.

Ellen Ochoa (b. 1958)

It is rare to hear Ellen Ochoa's name without 'the first Hispanic woman in space' following soon thereafter, a ritual incantation that serves less as a comma, taking us on to the rest of her life's achievements, but as a full stop, informing us that we have heard, in fact, all we need to know. Some go on to point out that she was also the first person of Hispanic descent, and only second woman, to act as director of the Johnson Space Center, NASA's nerve centre for human spaceflight operations, serving from 2013 to 2018. (The first woman director of Johnson, by the by, was Carolyn Huntoon, who served from Jan 1994 to Aug 1995.) That, however, is the firm limit, we are told, of what we need to know about Ochoa.

Well, nonsense to that. Ochoa is not just an astronaut, not just a long-serving and key administrator who saw JSC through some turbulent financial waters while updating practices and approaches that had grown decades old, but a really cool scientist too who got her bachelor's degree in physics from SDSU before moving on to graduate degrees in electrical engineering from Stanford University. Her speciality was optical systems, and how to optimise them to automatically process information. This was obviously something that could be of use in space exploration, where lags in communication between satellites and Earth makes automated optical observation systems an appealing notion. She is the holder of a number of patents related to object recognition and signal noise removal, both key areas in our continued and deeper investigation of the cosmos around us. For some time, she held the impressive title of Chief of the Intelligent Systems Technology Branch at NASA Ames, which was further devoted to the task of giving machines a more intelligent approach to noticing and recording objects of interest. While working on optical systems, she applied to the astronaut programme in 1985, then again in 1987, and was finally accepted in 1990, paving the way for her first spaceflight as a mission specialist on STS-56 in 1993, and for three subsequent spaceflights thereafter, the last being STS-110 in 2001.

Maria Zuber (b. 1958)

The descendant of Pennsylvania coal miners who perished at the hands of the black lung, planetary scientist Maria Zuber is today the co-chair of the President's Council of Advisers and Technology for the Joe Biden administration, and one of the most respected scientists of her generation. As the principal investigator for the 2011 Gravity Recovery and Interior Laboratory (GRAIL) lunar science project, she was the first woman to head a NASA spacecraft mission. That project launched two vehicles, called Ebb and Flow as the result of a national classroom naming contest, towards the Moon. Each of these was able to precisely measure the position of the other, and by recording small changes in those positions as a result of the Moon's gravity, GRAIL was able to produce a gravitational map that could then be used to more accurately determine its interior structure. Zuber has been a leading force for the last three decades in mapping the residents of our solar system, including membership on ten different NASA planetary missions, and her expertise has guided the high-resolution laser-altimetry mapping of Mars and the Moon as well as other mapping missions of Mercury and various large asteroids, using gravitational and topographical information to spur insight into planetary and lunar interiors.

Peggy Whitson (b. 1960)

You do not need me to tell you about Peggy Whitson, but I am going to do it anyway. Here in the United States she has attained a Lunchbox Hero status approaching that of Sally Ride in recognition of her long slate of spaceflight accomplishments. Her 665 days in space are the longest of any NASA astronaut, ever (the all-time record, in case you are wondering, is 803 days, held by cosmonaut Gennady Padalka). She holds the fifth all-time position on time spent on EVA missions, and the all-time record for number of spacewalks performed by a woman. She is the oldest woman ever to perform a spacewalk, and on her final flight became the oldest woman ever to orbit the Earth. She is the only woman who ever twice commanded the International Space Station, the first woman to serve as Chief Astronaut and, until 2019, held the record for the longest consecutive number of days spent by a woman in space, at 289. Little wonder, then, her classroom status as a role model for endurance and perseverance, but what tends to get lost in that barrage of global-level achievement is her level of scientific accomplishment.

She earned a doctorate in biochemistry from Rice University in 1986, where she returned as an adjunct assistant professor in Biochemical and Genetic Engineering in 1997, and served from 1992 to 1995 as a project scientist for the Shuttle-Mir Program. And the story is not over yet! The privately funded Axiom Mission 2, which with any luck will have launched by the time of this book's publication, will feature Whitson's return to space as mission commander, let us hope one of many to come.

Kalpana Chawla (1962–2003)

For a certain generation of space enthusiasts, the mention of Kalpana Chawla's name is as sure to evoke involuntary tears as the mention of Judith Resnik or Sharon McAuliffe's name continue to produce for those brought up a couple of decades earlier. It is difficult not to let the tragedy of her end, as one of the seven astronauts killed during the Columbia disaster, become the salient detail of her life, so emotionally impactful was that event for space travel lovers the world over, but that would not be doing fair service to Chawla's extraordinary career up to that point. A flight enthusiast since childhood, Chawla studied aeronautical engineering in her native India before going to the United States to do her master's work in aerospace engineering, beginning her association with NASA in 1988, and three years later becoming a naturalised citizen. Prior to applying to the astronaut programme, she carried out fluid dynamics research for vertical/short take-off conceptual models, and retained Commercial Pilot and Flight Instructor status across several different types of aircraft. In 1991, she applied for the astronaut programme, and joined in 1995. Whereas many astronauts have to wait upwards of half a decade for their first flight after joining the programme, Chawla was fortunate to have her first flight, STS-87, in 1997, the flight that launched her into history as the first Indian woman in space. The ship that carried her on her 252 revolutions of the Earth was the *Columbia*.

For four years following her historic first flight, Chawla completed technical work for the space station project while waiting for her next flight assignment, which duly came in 2001. That ill-starred mission launched in 2003, and shortly after take-off some foam insulation struck the port wing of the shuttle, doing just enough damage to the heat shielding to allow the super-heated gases produced on the shuttle's re-entry into atmosphere to penetrate the wing, destroy the craft, and kill all of the crew inside. Today, there are streets and fictional space vehicles and even a hill on Mars named after her, testaments from individual Earthlings of all walks of life to an individual who represented for them the best of our spirit of curiosity and exploration.

Lisa Randall (b. 1962)

Somerville. Clerke. Hogg. Randall.

It is The Lineage – a 200-year tradition of astronomy communication that has kept the population of two continents and beyond steadily informed about, and awestruck by, the universe above and around them. For a solid decade, stretching from 2005's *Warped Passages: Unraveling the Mysteries of the Universe's Hidden Dimensions* to 2015's *Dark Matter and the Dinosaurs: The Astounding Interconnectedness of the Universe*, Lisa Randall was the person we turned to when we needed the increasingly strange world of modern astronomy explained to us, with its supersymmetries, dark matter, and disturbing preponderance of baryons. Once initiated, we were able to follow her further, and appreciate her not only as an author, but as one of the theorists behind the Randall-Sundrum models, which sought to answer the age-old question: Why is gravity so weak, relatively speaking? It is 10^{24} times weaker than the force that we *explicitly named* The Weak Force. What gives?

The R-S models explain this using a braneworld system of Randall and Sundrum's own construction. Braneworld theories posit that our familiar reality is the result of our restriction to a three-dimensional system (called a 'brane') that is lodged within and interacts with a larger dimensional structure (called a 'bulk'). Under braneworld type systems, then, the 'powerful' fundamental forces (the electromagnetic, weak and strong) have such a pronounced magnitude because they are native to, and stuck in, our brane, while the 'weaker' ones (like gravity) have their homes in the wide expanses of the bulk, and our experience of them is only as a shadowy interaction, or bleed, between the bulk and the brane. The R-S models, published in 1999, posited two braneworld models that required only one fine tuning, as opposed to the two in previous theories: RS-1, a highly warped finite five-dimensional bulk, in which two branes exist, one of pretty strong gravitational force, and our own with a pretty weak gravitational force, with all the basic particles except the graviton localised to the branes, and RS-2, an infinite five-dimensional bulk with only one brane.

All of this is tremendously cool to contemplate, and also happens to be a wonderful setup for the beautifully terrible joke:

> 'Why do zombie particle physicists prefer RS-1 to RS-2?'
> 'Because it has twice as many branes.'

Andrea M. Ghez (b. 1965)

Recipient of the 2020 Nobel Prize in Physics, Andrea Ghez is perhaps the widest known of those astronomers born in the second half of the century that we shall be covering, and for good reason. From her position as head of the Galactic Center Orbits Initiative at UCLA, she pushed for the advancement from computationally heavy methods of reducing atmospheric distortion to new laser-guided adaptive optics systems at the massive Keck telescopes in Hawaii that allowed the tracking of star motion near the centre of our galaxy, and then used those orbits to prove the existence of a supermassive black hole, Sagittarius A*, lying in the core of the Milky Way. With masses some hundreds of thousands of times greater than ordinary black holes, supermassive black holes drive some very extreme processes in the universe, and are thought to reside at the centre of most galaxies. By using infrared frequencies, which are something on the order of a million times better at making their way through galactic dust than optical frequencies, and adaptive optics, which uses the distorting of a light-gathering mirror to compensate for atmospheric interference, Ghez was able to obtain not only motion data on stars near the galactic core but spectroscopic data that permitted her to compute their three-dimensional orbital motion. Those orbits allowed her to demonstrate that there must be a mass some 4 million times that of our Sun located at the centre of the Milky Way – in other words, a supermassive black hole. It was a tour de force of evolving observational techniques deployed over a patient quarter-century of time, and a powerful confirmation of astrophysical theory, that in every way deserved not only the Nobel Prize but the flood of admiration unto adulation that followed in its wake.

Conny Aerts (b. 1966)

Aerts is a pioneer in the relatively new field of asteroseismology, which promises to create vastly improved models of stellar interiors than those we have been drudging along with for the past century. By using the fluctuations in brightness of a star as mapped over the course of years by space telescopes like Kepler, asteroseismologists are able to obtain the many different oscillation modes of a stellar surface, the vibrational states that combine to produce the observed rhythmic variation over time. The pattern of those oscillations, or star quakes, then, can be related to what must be happening in the interior of the star – what kind of chemical mixing is happening? What is the rate of convection? How strong are the interior magnetic fields? Aerts' analysis of stellar seismic activity has led to a number of discoveries about star interiors, from her groundbreaking use of pulsations to describe a massive star interior for her dissertation in 1993 to her discovery of non-rigid rotation of a massive star's interior in 2003 to her work of the 2010s, opening stellar interiors thanks to the mass of Kepler data that began pouring out in the late 2000s. For her revolutionary work, Aerts has received a number of awards, including the Kavli Prize (2022) and the Francqui Prize (2012).

Lisa Kewley (b. 1974)

Our universe, by our best current reckoning, is some 13.8 billion years old. The lifespan of the stars that comprise that universe, and which made their arrival on the scene some 13.7 billion years ago, varies from 10 million to 10 billion years, meaning that we currently live at a point where generations of stars, and the galaxies they comprise, have risen and fallen. That brings up an interesting question – is there a way to tell, just looking at a galaxy, which generation it might belong to? Finding tell-tale chemical markers that answer that question, and therefore tell us information about not only what galaxies used to be like, but what we can expect from future galaxies, has been a major component of Lisa Kewley's highly decorated career. In 2005, she won the Annie Jump Cannon Award for her studies of oxygen content in galaxies of different ages, stretching back nearly 10 billion years in scope, and has since expanded her programme to investigate the metallicity (presence of atoms that are not hydrogen or helium), pressure variations, electron density measures, and ionisation parameters (which measure the abundance of ionising photons as against the density of the available hydrogen they are trying to ionise) of different galaxies, with an eye towards how those factors impact galactic evolution and star formation rates.

Feryal Özel (b. 1975)

In 2020, humans had their first opportunity to look at an actual picture of a black hole, courtesy of the Event Horizon Telescope of which Feryal Özel served as the modelling lead scientist. It was a profound moment that, for many, allowed us to experience that strange snap of space and time that comes when we realise how large the distance is between experiencing an object within its familiar theoretical stomping grounds, and experiencing it as a living piece of tangible reality, as if a stegosaurus moved into the house next door. For Özel, it was a moment among many spent in the company of the universe's most extreme objects, attempting to get at the physics behind early galaxy formation, black holes, magnetars (neutron stars with around 1,000 times more powerful magnetic fields than normal, which may result from the merger of two neutron stars from a binary pair) and neutron stars. Her projects for the future include using Doppler tomography (which, like other tomographic methods, involves stitching together multiple slices of observation over time) to map the surfaces of neutron stars and provide more detailed images of black hole event horizons.

Margaret Turnbull (b. 1975)

Before the Earth's oceans start vaporising some 1.8 billion years from now, humanity (or what is left of it/what race of super intelligent cats has succeeded it) is going to need to find a place to go. The problem is, there are a lot of stars out there, which are hard enough to get clean pictures of as it stands, let alone any potential planets orbiting them. Margaret Turnbull has been among those leading the charge in developing and perfecting new technologies to discover and categorise exoplanets using coronagraph technology like the Near Infrared Camera and Multi-Object Spectrometer on Hubble, the Near Infrared Camera on James Webb, or the Coronagraph Instrument scheduled to launch with the Nancy Grace Roman Space Telescope sometime in 2027. The idea of the stellar coronagraph is simple enough in concept: if you want to find small things near a star, just cover up the star and its immense glare, and look for objects passing in front of the still luminous but far less overwhelming (by a factor of about a million) corona, or outer layer. In addition to finding new exoplanets, Turnbull is a leader in the field of determining their potential habitability, and with Jill Tarter (see above) she is responsible for the HabCat project, which has identified thirty potentially habitable star systems within 100 light years of Earth.

Marta Burgay (b. 1976)

PSR J0737-3039. To most of us, that collection of letters and numbers resembles something similar to the restocking code for a brand of high-end soup. But to astronomers, there is magic in that jumble, for it is the designation of the universe's only known (so far) double pulsar. Pulsars, as we remember from our time with Jocelyn Bell Burnell, are magnetised rotating neutron stars that shoot electromagnetic radiation out of their magnetic poles, which manifest themselves to us as regular periodic bursts. PSR J0737-3039 was discovered by Marta Burgay and her team in 2003, and observation of the double pulsar over the last two decades has been used to confirm with stunning precision some of general relativity's predictions about gravitational waves' impact on orbital periods. Since that observation, she has continued her work in using binary pulsars (systems consisting of a pulsar locked in orbit with either a neutron star or a white dwarf) to further test the gravitational predictions of relativity theory.

Alicia M. Soderberg (b. 1977)

Soderberg is a specialist in supernova forensics, and if there is a cooler thing to have on your business card, I cannot imagine what it might be. Her speciality is the death of stars, and, in particular, probing supernovae for non-optical emissions. Usually, astronomers only get to observe supernovae some weeks after they have exploded, and must rest content with mining what information they can from the remaining measurable radiation. In 2006, however, Soderberg was fortunate enough to observe a supernova in the act of exploding, and was able to capture gamma ray signals from that moment of spectacular collapse. Her research into the link between supernovae and gamma ray production won her the Annie Jump Cannon Award in 2009. Her work of the last decade has focused on uncovering and analysing new classes of supernovae using data gathered across the electromagnetic spectrum, including long wavelength information from the Expanded Very Large Array (EVLA) and Atacama Large Millimeter Array (ALMA).

Thushara Pillai (b. 1980)

At a distance of 1,400 light years away there is a cosmic gas cloud in which stars are slowly forming called the Serpens South region. Discovered some fifteen years ago, it is an ideal location to closely observe the processes that govern the formation of stars from gas, and in particular to determine the relative contributions of gravity, turbulence and magnetism in stellar creation. Thushara Pillai, employing the HAWC+ camera system, tackled the question of magnetic fields in Dark Clouds, which required incredibly fine work using the polarisation of infrared light emitted by dust grains in the cloud to determine the direction and strength of the cloud's magnetic fields. From these observations, Pillai was able to draw parallels between the density of clouds and their orientation relative to the existing magnetic field, and also to form theories about the role that magnetic fields play in the formation of high mass stars.

Megan Lin

Since 2021, Lin has been the Project Manager for the Curiosity Mars Science Laboratory Rover, the culmination of a decade of mechanical engineering work on Curiosity. While still an undergraduate at Northeastern University, she began working at the famous Jet Propulsion Laboratory in 2008, on developing more efficient methods of drilling and core sampling which were then incorporated into the project. In November 2011, Curiosity was launched, and Lin, who received her bachelor's in 2010 and would finish her master's at UCLA by 2013, worked as a mechanical engineer on the sample acquisition and handling group, while developing new drilling techniques and tools for future rover missions. She then added a whole separate job to her plate by becoming Project Manager for the Airborne Snow Observatory, which over four years required her to spend 300 hours in the back of a plane taking snow pack measurements for the project, in addition to her engineering and administrative duties. In 2020, she was made Deputy Project Manager of Curiosity, and advanced to full Project

Manager by 2021. As of this writing, Curiosity is on the ninth year of its originally scheduled two-year mission, and has explored 17 miles of the Martian surface.

Burçin Mutlu-Pakdil

Burçin's Galaxy is one of the most fantastic things we know of in space, a double-ringed elliptical galaxy consisting of a young outer ring (a mere 130 million years old) and a faint red inner ring of different composition and unknown age, all surrounding a nearly perfectly round old inner core (around 5.5 billion years old). It is a thoroughly one-of-a-kind celestial object, and it was discovered by the one-of-a-kind Burçin Mutlu-Pakdil during her time as a graduate student. Mutlu-Pakdil grew up in Turkey, and in 2009 became the first of her family to get a college degree. She subsequently received her PhD from the University of Minnesota, Twin Cities, in 2017 for her dissertation on supermassive black hole scaling relations. Her research since then has focused on survey programmes centred on 'peculiar' objects, including dwarf galaxies, dark matter halos (and how they might give rise to stars), exotic ringed galaxies, and black holes. She is also an advocate for greater representation of Muslim women in science.

Yaël Nazé (b. 1976)

Sometimes called the Belgian Carl Sagan, Nazé is both an astrophysicist, specialising in the analysis of massive stars, and a science communicator who has written multiple books in the French language about the history and practice of astronomy (including the volume *Femmes Astronomes* which, if you liked this book and read French, you will probably enjoy even more) as well as a series of articles that answer questions everybody has had at one point or another – 'If the centre of the Earth is as hot as the Sun, why doesn't the Earth burn?', 'Can people live on Saturn?', and 'If the Sun warms the Earth, why is it cold in space?' among others. Her research focuses on the role that large stars play as the monarchs of the universe, in particular the part that magnetic fields play in their evolution, the nature of their X-ray emissions, the role that their solar winds play in the galactic ecology, what happens when two solar winds meet, and their unique chemical compositions.

Knicole Colon (b. 1986)

Exoplanet astronomer Knicole Colon has spent the better part of the last decade working on two of the most exciting scientific endeavours of the last half century: the Hubble Space Telescope and the James Webb Space Telescope. A native of New Jersey, she received her master's degree in 2012 from the University of Florida, and joined the Kepler project (which sought to discover Earth-like planets in other solar systems) at NASA Ames in 2015. That project lasted from 2009 to 2018, when Kepler at last ran out of fuel, but just before its end, in 2017, Colon moved on to the Goddard Space Flight Center to start working on the Hubble while continuing her support of the Kepler mission. Whether working on Kepler, or Hubble, or KELT (the marvellously named Kilodegree Extremely Little Telescope), or TESS (the less marvellously named but still thoroughly excellent Transiting Exoplanet Survey Satellite), or in her newest position as Deputy Project Scientist for Exoplanet Science on the James Webb Space Telescope, her interest has consistently been the study of planets outside our solar system, with a particular focus on 'extreme' exoplanets that do not resemble any of those in our system and which therefore have new tales to tell about planet formation and evolution.

Selected Reading List

General Volumes

Ackman, Martha, *The Mercury 13: The True Story of Thirteen Women and the Dream of Space Flight.* (Random House, 2003).
Bernardi, Garbriella, *The Unforgotten Sisters: Female Astronomers and Scientists Before Caroline Herschel.* (Springer Praxis, 2016).
Böhm-Vitense, Erika, *Introduction to Stellar Astrophysics.* (Cambridge, 1989).
Brück, Mary, *Women in Early British and Irish Astronomy: Stars and Satellites.* (Springer, 2009).
Carroll, Bradley W. and Dale A. Ostlie, *An Introduction to Modern Astrophysics.* (Cambridge 2017).
Druyan, Ann, *Cosmos: Possible Worlds.* (National Geographic, 2020).
Gibson, Karen Bush, *Women in Space: 23 Stories of First Flights, Scientific Missions, and Gravity-Breaking Adventures.* (Chicago Review Press, 2020).
Haines, Catharine M.C., *International Women in Science: A Biographical Dictionary to 1950.* (ABC-Clio, 2001).
Hogg, Helen Sawyer, *The Stars Belong to Everyone: How to Enjoy Astronomy.* (Doubleday, 1976).
Jenkins, Dennis, *Space Shuttle: The History of Developing the National Space Transportation System, The Beginning Through STS-75.* (Motorbooks, 1996).
Kevles, Bettyann Holtzmann, *Almost Heaven: The Story of Women in Space.* (Basic Books, 2003).
Kozai, Yoshihide, and Seeichiro Kawashima, Takeshi Tominaga, Tomoko Hisatome, and Katsuko Saruhashi, ed., *My Life: Twenty Japanese Women Scientists* (Uchida Rokakuho, 2001).
McGrayne, Sharon Bertsch, *Nobel Prize Women in Science: Their Lives, Struggles, and Momentous Discoveries.* (Joseph Henry Press, 1993).
Nazé, Yaël, *Femmes Astronomes*, (Biblis, 2014).

Ogilvie, Marilyn Bailey, *Women in Science: Antiquity Through the Nineteenth Century: A Biographical Dictionary with Annotated Bibliography.* (MIT Press, 1986).
Ogilvie, Marilyn Bailey and Joy Harvey, ed., *The Biographical Dictionary of Women in Science: Pioneering Lives from Ancient Times to the Mid-20th Century.* 2 vols. (Routledge, 2000).
Peterson, Barbara Bennett, ed., *Notable Women of China: Shang Dynasty to the Early Twentieth Century.* (Routledge, 2015).
Randall, Lisa, *Dark Matter and the Dinosaurs: The Astounding Interconnectedness of the Universe.* (Ecco, 2015).
Rossiter, Margaret W., *Women Scientists in America: Volume 1: Struggles and Strategies to 1940.* (Johns Hopkins, 1982).
Schiebinger, Londa, *Schöne Geister: Frauen in den Anfängen der modernen Wissenschaft.* (Klett-Cotta, 1993).
Shetterley, Margot Lee, *Hidden Figures.* (William Morrow, 2016).
Sobel, Dava, *The Glass Universe: How the Ladies of the Harvard Observatory Took the Measure of the Stars.* (Viking, 2016).
Strohmeier, Renate, *Lexikon der Naturwissenschaftlerinnen und naturkundigen Frauen Europas.* (Verlag Harri Deutsch, 1998).
Weitekamp, Margaret A., *Right Stuff, Wrong Sex: America's First Women in Space Program.* (Johns Hopkins University, 2004).
Yost, Edna, *Women of Modern Science.* (Dodd, Mead & Co., 1959).
Zimmerman, Robert, *The Universe in a Mirror: The Saga of the Hubble Space Telescope and the Visionaries Who Built It.* (Princeton University Press, 2008).

Memoirs and Biographies

Becker, Barbara J., *Unravelling Starlight: William and Margaret Huggins and the Rise of the New Astronomy.* (Cambridge University Press, 2011).
Bell Burnell, Jocelyn, *A Quaker Astronomer Reflects: Can a Scientist Also be Religious?* (2013).
Bergland, Renee, *Maria Mitchell and the Sexing of Science.* (Beacon Press, 2008).
Brock, Claire, *The Comet Sweeper: Caroline Herschel's Astronomical Ambition.* (Icon Books, 2007).
Brück, Mary, *Agnes Mary Clerke & the Rise of Astrophysics.* (Cambridge, 2002).

Cannon, Annie Jump and Doug West, *In the Footsteps of Columbus.* (2015).

Cochet, Jean Benoît Désiré, *Notice Biographique sur M. Nell de Bréauté, Correspondant de l'Institut* (1855).

Daugherty, Tracy, *Dante and the Early Astronomer: Science, Adventure, and a Victorian Woman Who Opened the Heavens.* (Yale University Press, 2019).

Dyson, Marianne J., *A Passion for Space: Adventures of a Pioneering Female NASA Flight Controller* (Springer, 2016).

Goss, W.M., *Making Waves: The Story of Ruby Payne-Scott: Australian Pioneer Radio Astronomer.* (Springer, 2013).

Hill, Edward, *My Daughter Beatrice: A Personal Memoir of Dr Beatrice Tinsley, Astronomer.* (American Physical Society, 1986).

Hoskin, Michael, *Caroline Herschel: Priestess of the New Heavens.* (Science History Publications, 2013).

Jemison, Mae, *Find Where the Wind Goes: Moments from My Life.* (Signal Hill Road, 2020).

Johnson, George, *Miss Leavitt's Stars: The Untold Story of the Woman Who Discovered How to Measure the Universe.* (W.W. Norton, 2005).

Lemonick, Michael D., *The Georgian Star: How William and Caroline Herschel Revolutionized Our Understanding of the Cosmos.* (W.W. Norton, 2009).

MacPike, Eugene Fairfield, *Hevelius, Flamsteed and Halley: Three Contemporary Astronomers and their Mutual Relations.* (Taylor and Francis, 1937).

Mitton, Jacqueline and Simon Mitton, *Vera Rubin: A Life.* (Belknap Press, 2021).

Moore, Donovan, *What Stars are Made of: The Life of Cecilia Payne-Gaposchkin.* (Harvard University, 2020).

Payne-Gaposchkin, Cecilia, *Cecilia Payne-Gaposchkin: An Autobiography and Other Recollections.* (Cambridge University, 1996).

Rubin, Vera, *Bright Galaxies Dark Matters.* (American Institute of Physics, 1997).

Scoles, Sarah, *Making Contact: Jill Tarter and the Search for Extraterrestrial Intelligence.* (Pegasus Books, 2017).

Seddon, Rhea, *Go For Orbit: One of America's First Women Astronauts Finds Her Space.* (Your Space Press, 2015).

Sherr, Lynn, *Sally Ride: America's First Woman in Space.* (Simon & Schuster, 2014).

Sharpe, Mitchell R., *'It is I, Sea Gull': Valentina Tereshkova, First Woman in Space.* (Thomas Y. Crowell Company, 1975).

Selected Reading List

Sullivan, Kathryn D., *Handprints on Hubble: An Astronaut's Story of Invention.* (MIT Press, 2019).
Teitel, Amy Shira, *Fighting for Space: Two Pilots and their History Battle for Female Spaceflight.* (Grand Central Publishing, 2020).
Walz, Eric, *Die Sternjägerin.* (Blanvalet Taschenbuch Verlag, 2009).
Westphal, Johann Heinrich, *Leben, Studien und Schriften des Astronomen Johannes Hevelius.* (Konigsberg, 1820).

Index

A
Aerts, Conny, 214
Aglaonike, 4
Albertson, Mary Ann, 67
Al 'Ijliyyah, 4
Aryabhata, 22

B
Barney, Ida, 184–5
Berlin Academy of Sciences, 12–14, 32
Berlin Observatory, 12–14, 32
Blagg, Mary Adela, 73–4
Böhm-Vitense, Erika, 200–201
Bond, Selina, 72
Brahe, Sophie, 25–6
Brahe, Tycho, 25–6
Bréauté, Eléonore, 41
British Astronomical Association, 66, 68, 176, 180–1
Brown, Elizabeth, 66
Bryan, Margaret, 38-9
Burbridge, Margaret, 102–107
Burgay, Marta, 216
Burnell, Jocelyn Bell, 126–32, 216
Byrd, Mary Emma, 70–1

C
Cannon, Annie Jump, 56, 77, 80–2, 84, 184
Chawla, Kalpana, 211–12
Clarke, Edward, 44

Clerke, Agnes Mary, 50–3, 66
Cobb, Jerrie, 139–43
Cochran, Jacqueline, 140–3
Cohen, Judith Gamora, 204
Colon, Knicole, 219
Cook, Alice Grace, 175–6
Copernicus, Nicolaus, 10
Cunitz, Maria, 6–9
Cunningham, Susan Jane, 54, 68–9
Cushman, Florence, 74

D
Dark matter, 114–15
Draper Catalogue, 56, 59–61, 74–5, 77, 84, 177–8
Dumée, Jeanne, 28
Dyson, Marianne, 151–6

E
Eddington, Arthur, 83
Eimmart, Marie Claire, 30–1
Enheduanna, 3
Evershed, Mary Acworth, 178–9

F
Faber, Sandra, 203–204
Fallows, Mary Anne, 37–8
Farnsworth, Alice, 183, 187
Fatima dea Madrid, 4–5
Flamsteed, Margaret, 28–9
Fleming, Williamina, 57–61
Freundlich, Erwin, 195–6

Index

Funk, Wally, 143
Furness, Caroline, 181–2

G
Galilei, Galileo, 63, 99
Gaze, Vera Fedorovna, 191
Ghez, Andrea, 213
Gill, Jocelyn Ruth, 197
Gökdoğan, Nüzhet, 195
Gorelick, Sarah, 143

H
Haigneré, Claudie, 207
Hansen, Julie, 185
Harlay, Marie-Jeanne, 39–40
Hart, Jane, 142
Harvard Observatory, 57–61, 68, 70, 72–4, 77–9, 81, 84–6, 176–7, 182, 189, 191, 194
Harwood, Margaret, 183–4
Hasan, Hashima, 205–206
Heger, Mary Lea, 188–9
Herschel, Caroline, 15–20
Herschel, William, 16–18, 75
Hevelius, Elisabetha, 26–8
Hevelius, Johannes, 27
Hixson, Jean, 144
Hoffleit, DOrrit, 184, 193–4
Hogg, Helen Sawyer, 50, 88–91, 182–3
Hoyle, Fred, 102, 105, 127
Hubble, Edwin, 102
Hubble Space Telescope, 104, 106, 120–5, 162, 165–6, 203, 205, 219
Huggins, Margaret, 46–9, 66
Huggins, William, 46–9

I
Iwanowska, Wilhelmina, 192

J
James Webb Space Telescope, 206, 215, 219
Janssen, Henrietta, 71
Jemison, Mae, 171–4
Jewitt, David, 133–5

K
Karachkina, Lyudmila, 204–205
Keck Telescope, 203–204, 206, 213
Kepler, Johannes, 6, 10
Kewley, Lisa, 214
King, Susan Raymond, 186–7
Kirch, Christine and Margaretha, 31–3
Klumpke-Roberts, Dorothea, 75–6
Kondakova, Elena, 207–208
Koyama, Hisako, 98–101

L
Lalande, Joseph-Jérôme, 34–7, 39–40
Leavitt, Henrietta Swan, 77–9, 84, 184
Leibniz, Gottfried Wilhelm von, 12
Leland, Evelyn Frances, 182
Lepaute, Nicole-Reine Hortense, 33–5
Leverton, Irene, 143
Leviathan, 64–5
Lick Observatory, 189, 203
Lin, Megan, 217–18
Llewelyn, Thereza, 66–7
Luu, Jane, 133–5

M
Makemson, Maud, 186
Manfredi, Maddalena and Teresa, 29–30

Massevich, Alla Genrikhovna, 198–9
Maunder, Anne, 100, 179–81
Maury, Antonia, 176–7
Mei Wending, 23
Menzel, Donald, 194
Mercury 13, 138–44
Mitchell, Maria, 42–5, 183–4
Mount Holyoke, 182, 187
Mutlu-Pakdil, Burçin, 218

N
Nagaraja, Mamta Patel, 168–70
NASA 117-18, 121–5, 138–44, 151–5, 157–61, 163–6, 168–70, 173–4, 186, 197, 199, 205–206, 209–12, 219
Nazé, Yaël, 218

O
OBAFGKM system, 59–60, 81
Ocampo, Adriana, 206
Ochoa, Ellen, 209
Oterma, Liisi, 196–7
Otté, Elise, 62–3
Özel, Feryal, 215

P
Parsons, Mary, 64–5
Payne-Gaposchkin, Cecilia, 77, 83–7, 105, 189, 191
Payne-Scott, Ruby, 92–7
Pettit, Hannah Steele, 185
Pickering, Edward Charles, 55, 57–60, 70, 72, 77–9, 81, 177–8, 182
Pierry, Louise Elizabeth de, 35–7
Pillai, Thushara, 217

Pişmiş, Paris, 195–6
Pogson, Isis, 66–8
Prince, Helen Dodson, 193
Pulsars, 129–31, 216

R
Radcliffe College, 81, 88, 184, 188–9, 191
Radiophysics Laboratory, 94–6
Randall, Lisa, 212–13
Ride, Sally, 157–61
Roman, Nancy Grace, 104, 106, 120–5
Royal Astronomical Society, 66, 68, 74, 181, 182, 185, 190
Rubin, Vera, 112–15
Russell, Henry Norris, 85, 190

S
Sabine, Elizabeth, 62–3
Savitskaya, Svetlana, 207–208
Scarpellini, Caterina, 63–4
Seondeok, 1–2
Serova, Yelena, 208
SETI Program, 116–19
Shajn, Pelageya, 188
Shapley, Harlow, 85, 88, 189, 191, 194
Short, Maria, 40–1
Shuhua, Ye, 201–202
Sitterly, Charlotte, 190
Sloan, Jerri, 143
Soderberg, Alicia, 216
Somerville, Mary, 50
Sputnik, 138, 198
Steadman, Bernice, 143
Stumbough, Gene, 144
Sullivan, Kathryn, 162–7
Swope, Henrietta Hill, 191, 193

Index

T
Tarter, Jill, 116–19
Tereshkova, Valentina, 145–50, 207
Terian, Alenush, 199–200
Tinsley, Beatrice, 108–11
Tseraskaya, Lidiya, 71
Turnbull, Margaret, 215–16

U
Underhill, Anne Barbara, 199
Universal Time, 201–202

V
Vassar College, 44, 54, 69–70, 181
Vostok, 145, 147–8
Vyssotsky, Emma, 188

W
Wang Zhenyi, 21–4
Ward, Mary, 65
Wellesley, 54–5, 81
Wells, Agnes, 183
Whiting, Sarah Frances, 54–6
Whitney, Mary Watson, 69–70, 181
Whitson, Peggy, 210–11
Wilson, Fiammetta, 175–6
Winkelmann, Maria, 10–14
Winlock, Anna, 72
Woltman, Rhea, 144
Wright, Frances, 189

X
Xu Guangqi, 23

Y
Yerkes Observatory, 50, 104, 106, 121, 185, 187, 196–7
Young, Ann Sewell, 182–3

Z
Zhuravleva, Lyudmila, 204–5
Zuber, Maria, 210